D0246616

CITY ABLAZE

Life with the World's Busiest Fire-fighters

CITY ABLAZE

Life with the World's Busiest Fire-fighters

MARTIN LLOYD-ELLIOTT

B L O O M S B U R Y

This book is dedicated to the firemen and firewomen
of Soho, past, present and future. Thank God
for them, for they are the real cavalry!

Author's Note

I believe all the stories in this book are true. They are based principally on the transcripts of interviews between me and dozens of retired and serving members of the London Fire Brigade. Wherever practical, stories have been corroborated, but on occasion this has not been possible. I include them in the text in good faith. Unless otherwise stated, all the opinions expressed in this book are my own, and not the opinions of the London Fire and Civil Defence Authority.

First published in Great Britain 1992
Bloomsbury Publishing Limited,
2 Soho Square, London W1V 5DE

Copyright © 1992 by Martin Lloyd-Elliott

The moral right of the author has been asserted

PICTURE SOURCES

Chartered Insurance Institute: page 34
Daily Mail: page 177
Harry Errington, GC: page 58
Sally Holloway: page 37
Independent/Herbie Knott: page 184
Independent/Jeremy Nicholl: page 183
Martin Lloyd-Elliott: pages 2, 10, 14, 86, 90, 103, 107, 118, 121, 132, 166, 186, 187, 188, 191, 193, 195, 198
London Fire Brigade: pages 44, 45, 48–9, 52, 61, 66, 67, 70, 71, 72, 74, 75, 78, 79, 80, 81, 83, 95, 96, 97, 114, 154, 155, 156, 164, 169, 174
London Fire Brigade/M.R.L.: pages 110, 111
London Fire Brigade/B.G.M.: page 113
London Fire Brigade/D.A.P.: page 91
London Fire Brigade/M.J.T.: page 94
National Gallery of Victoria, Australia (Felton Bequest): page 22
Gerald Paul: pages 99, 151
Martin Powell: page 170
Owen Rowland: pages 104, 117, 123, 143, 160, 161
S & G Press Agency Limited: pages 51, 128
SI Vistas: page 42
Today: page 182
Jonathan Waights: pages 29, 146
Westminster City Archives: pages 76, 77

Map on page 13 reproduced by permission of Geographers' A–Z Map Company Limited © Crown Copyright

Author photo on jacket flap: Martin Asser

A CIP catalogue record for this book is available from the British Library

ISBN 0-7475-1156-X

Designed by Bradbury and Williams
Typeset by Hewer Text Composition Services, Edinburgh
Printed and bound by Butler and Tanner Limited, Frome and London

Contents

Greater love hath no man this, that
a man lay down his life for his friends.

John 15:13

Acknowledgements

To name and thank every person who has contributed to the creation of this book is impossible: dozens of firemen and firewomen of the London Fire Brigade have between them related literally hundreds of stories to me about their adventures and it has been a heart-rending task to leave out many wonderful stories and include so few. To everyone who has helped, whether named in the text or not, I say thank you with all my heart.

I must mention the following members of the London Fire Brigade who have been of particular assistance: Louisa Steel and the staff of the London Fire Brigade Library, Non-Operational Sub-Officer John Rodwell, Curator London Fire Brigade Museum and Non-Operational Leading Fireman Roy Still, Assistant Curator London Fire Brigade Museum, the staff of the Command and Mobilizing Centre and Non-Operational Sub-Officers Don Pye and Brian Mercer, London Fire Brigade Photographic Services. For his endless enthusiasm, information, inspiration, energy and technical advice, great thanks to ACO (Retd) Roy Baldwin. Thanks also to ACO Robin Graham, ADO Richard Nichol and Station Officer Roy Davies at the Fire Services College, Moreton-in-Marsh, for their assistance with the front cover photograph and Leading Fireman Roger Kendall and Christopher Kendall for their help. Many thanks to the Chief Fire Officer of the London Fire Brigade Gerald Clarkson, the North Area Commander ACO Ritherdon and Soho Fire Station Commander Rob Molson for their support of this project.

I shall always be particularly indebted to the Station Officers at Soho who made me so welcome and who have individually contributed greatly to the contents of the book: Tony Wilmott, Val Hawes, John Peen, Colin Townsley, Gerry O'Neil, Shiner Wright, Barry Humphries, Ray Chilton, Chris Staynings, Dick Haigh, Roy Wilsher and especially Bruce Hoad for his invaluable research and support and 'Turk' Manning, whose tales were the original inspiration for this book.

It has been an honour to ride with all the firemen of Soho Fire Station and I owe thanks to every one of them for their kindness, care, fun, love of life and adventure. In a service where 'Mention no names' is an unofficial motto, I must break the rules and thank Alan Hart, Steve Davies, Steve Bell, John Norris, Harry Errington GC, Paul Grimwood, Wolf Goodman, Ron Morris, George Phillips, Terry Spindlow, Knocker White, Tony Cooper, Pat Talbot, Ernie Allday and Jessica Ireland for their time, ideas, enthusiasm and encouragement.

Thanks also to Catherine Reynolds for starting me off and pointing me in the right direction, the Chairman and members of the Soho Society, the diligent, brilliant staff of the Westminster City Archives, the authors Sally Holloway and Judith Summers, my agent Anne Dewe, and Penelope Hoare for introducing me to her, Owen Rowland for his hilarious stories and stunning photographs and Alan Rowland for printing them, Alison Walkington for reading the scripts, Carole Salmon for the title and especially Alexa de Ferranti, Penny Phillips and David Reynolds at Bloomsbury for believing in the book.

Lastly I thank Gordon White, Head of Public Relations, London Fire Brigade, who made this book possible.

INTRODUCTION

In that part of town where one's senses are bombarded with every whisper, at every turn, blink, sniff or breath, there is one sight above all others that makes the blood run hot and fast, or, for those that know better, run cold: the three fire engines from Soho Fire Station charging their way through the West End traffic, flashing blue lights like a thousand dancing jewels thrown in the air and dashed against passing windows and walls, the sirens blaring a hundred-trumpet-loud shout of alert, engines roaring like tanks into battle. They are the real cavalry, dazzling in their ride to the rescue. The spectacle is daring, delightful, deathly: the men aboard modern-day darlings to the victims of humanity's cruel fates. They bring in their wake courage, concern, brute brawn and mighty muscle, succour, skill, wise counsel and glorious common sense governed by calm in the face of calamity. Their purpose is unique: to save the lives and properties of their fellow men at any cost to themselves.

They are a breed apart, these warriors of the fire brigade. Their 'brotherhood' is international, their language distinctive and specific – fire, like music, speaks to those who would challenge its power with a biting tongue the same the world over. The men of the London Fire Brigade, perhaps more than any others, have a splendid, proud heritage and a worldwide reputation for courage, well-earned. They are the guardians of a uniquely complex metropolis, encompassing every form of structure made by man and offering the potential for disaster in any form imaginable on earth. Rivalry between stations has always been a healthy and potent source of high standards of performance within the brigade. Fierce loyalties develop amongst men who serve at one fire station for many years and they will swear that theirs is 'the best', be it Clapham, Sidcup, West Ham or Kensington. Daring to suggest that another station might be better provokes derision and passion. But mention the name 'Soho' to a fireman and the response is either fiery, 'Why do people go on and on about that station as if it was something special? Why does everyone call the Soho blokes the "glory boys"?' or more often, 'Well, Soho's different. It's . . . well, you know . . . special.'

London's fire-fighters have served many masters in their time: the insurance companies of old, the Metropolitan Board of Works, the London County Council (LCC) and the Greater London Council (GLC). At present they are controlled by the London Fire and Civil Defence Authority (LFCDA) which has responsibility for all of the Greater London area from Enfield in the north to Croydon in the south, and from Dagenham in the east to Hayes in the west. Six and a half thousand operational staff divided into four shifts known as 'watches' operate 114 fire stations twenty-four hours a day, 365 days of the year. London is divided into five areas for ease of administration and control - North West (G), North (A), North East (F), South East (E) and South West (H). Each division is almost a separate brigade, like one of five battalions in a large regiment. Each is identified by a letter prefix and each station has its own number. On average London fire stations are responsible for an area, called 'the ground', of between five and ten square miles; Soho Fire Station covers but one square mile.

The heart of London is not the City: that is its head. The heart is Soho. Soho Fire Station, now situated in Shaftesbury Avenue on the corner of Gerrard Place, has one of the smallest 'grounds' in the world, within which some of the most important buildings in London exist. Its boundaries are the river embankment, Green Park, Whitehall, the Aldwych, Holborn, Bloomsbury, Gower Street and Bond Street. Purists will argue that the area covered by Soho Fire Station covers a great deal more than Soho village itself, and they are right. A real Sohoite wouldn't dream of associating with anyone who lived north of Oxford Street, west of Regent Street, south of Shaftesbury Avenue or east of Charing Cross Road, but to the men of Soho Fire Station, positioned at the very centre of their square mile of territory, anywhere on their patch counts as Soho, and that is that!

The boundaries of the present 'ground' were defined in 1921 when the new station in Shaftesbury Avenue was opened, replacing three older stations. The station now has responsibility for dozens of theatres, cinemas, nightclubs, churches including St Martin-in-the-Fields, some of the greatest hotels in London including the Savoy and the Ritz, the British Museum, the National Gallery, the London Library, the Royal Academy of Arts, Clarence House and St James's Palace and some of the most famous street names in the world: Covent Garden, Leicester Square, Piccadilly Circus, the Strand, Oxford Street, the Haymarket, Horse Guards Parade, Trafalgar Square, the Mall. Its shops and services read like a who's who of

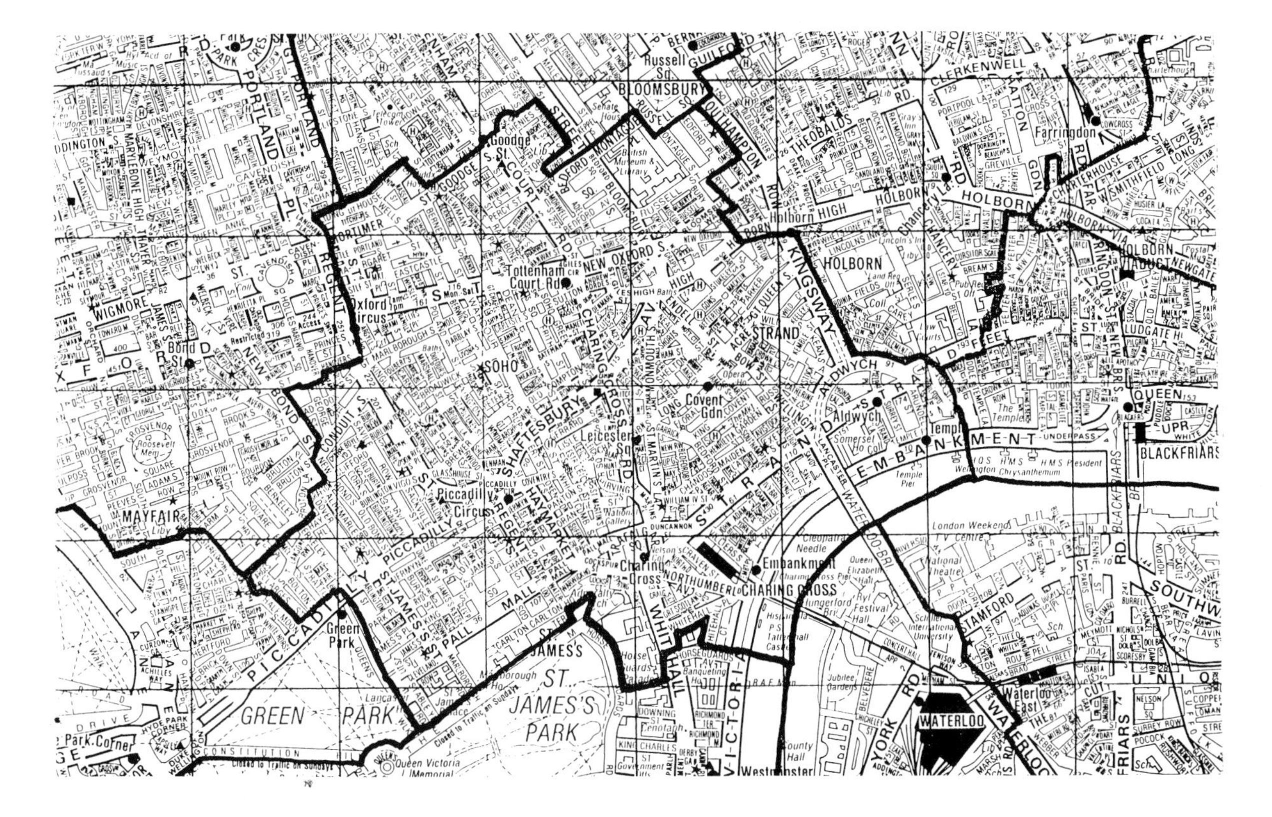

Map showing the area presently covered by Soho Fire Station.

everything elegant: Christie's and Spink's the auctioneers, the Edwardian splendours of the grand shops of Jermyn Street and Burlington Arcade, the gentlemen's clubs of Pall Mall and St James's. Then there is Soho itself: the sleazy, 'sexy' clubs, Paul Raymond's extensive property and entertainments empire, the heart of the British Film Industry, a large section of the rag trade, the last bastion of craftsmen, the creative skilled legendary masters of old: cabinet-makers, tailors, embroiderers, musical-instrument makers, designers, chefs, gilders . . . Anything you might want to purchase can be found here, from the finest ancient Chinese silks at Liberty's (and not as might be hoped in Chinatown) to the most extraordinary variety of toys your children ever dreamed of at Hamleys. The whole ground is a glittering cabaret of household names, synonymous the world over with the very heart of London town.

The population changes enormously during the day, rising from a low residential number of a few thousand to as many as a million people in that one square mile on a busy Christmas shopping day. Streets that are always deserted at four in the morning become jammed solid with traffic and crowds of people. On a Friday and Saturday night the pavements are thick with revellers, and a semi-permanent spirit of carnival reigns as tens of

Like a great London smog, thick smoke envelops the pumping engines as they feed water through miles of spaghetti-like hose to the men of Soho struggling to subdue an underground inferno.

thousands seek the bright lights for shows, films, clubs. Others squander money on drink, drugs and over-priced sexual diversions. Of these people, thousands walk past the front of Soho Fire Station oblivious of the large red doors until they whirr open to reveal the revving appliances preparing to burst out into the street. Most nights there are just twelve fire-fighters manning the station and they carry the enormous burden of protecting so many people. It is a thankless task, but one which, without exception, they relish.

In 1982 a brand new fire station was constructed in Shaftesbury Avenue. On 24 August 1983 one of the grand old men of British theatre, Sir Ralph Richardson, in what was to be his last public appearance before he died, addressed the men and officers of Soho Fire Station thus:

It's a great joy for me to be here! When I first came to London I opened at the Haymarket Theatre in 1921, the same year that the old Soho Fire Station opened in Shaftesbury Avenue. I've played in most of the theatres here in theatreland . . . There is a kindred spirit between my profession and yours. However we feel at any time when that curtain goes up, be it jolly or miserable, happy, well or ill, we've got to go on and make an effort.

You have no set time to your work and no time to get ready for your work as we have when they call the half hour. You may not have two minutes to get ready for it, when you go down that pole (I'd rather like to have a go at it myself). You never know what's going to happen. You might be out for half the night, or it might be a false alarm. That's one of the things we do escape in our job. I can't remember ever in my time the curtain going up and finding no single person in the theatre. You don't know what your job is to be: it might be the most perilous imaginable. And when you do go out, what a *dazzle* you make – how exciting it is, when those red doors open and the bells go, and out you come. What a thrill (unless you happen to be playing in the theatre next door, which we did do once . . .) I tell you this: there's not a man in the whole of London . . . who would be more proud than I am to take part in any celebration to do with the London Fire Brigade. Regarded with deep love, compassion, and wishing it great joy for its nobility – I'll always be proud!

With this he majestically walked across the appliance-room floor, commanding it as if he were once again treading the boards, till he reached the curtain-shrouded plaque. He grasped the cord, pulled it vigorously and declared the station open, accompanied by the ringing of the fire bell and the blaring of the sirens on the machines nearby; all the time his eyes twinkling, for he, like everyone who comes into contact with the firemen of Soho, had been touched by their magic.

The History of Soho's Firemen

oho takes its name from an ancient Anglo-French hunting call used when a hare was discovered. Hare-hunting was common in the fields north of the Oxford Road (now Oxford Street) which was the main route from the City of London to the gallows at Tyburn (now Marble Arch). The fields were almost all royal property and supported few buildings.

Apart from a small hamlet that had developed around the leper hospital at St Giles and also the royal mews at Charing Cross, the entire area was still rural farmland as late as 1540. During the next sixty years, as the City of London became overcrowded and grossly unhygienic, more and more residents began to build their homes in the area which was now known as Soho, and by the mid 1640s the fields around the parish church of St Martin-in-the-Fields were densely populated.

With the great plague in 1665 (which actually began in Drury Lane and subsequently spread to the City) and the Great Fire of London in 1666 (whose flames reached as far as Holborn), still more people moved to Soho, where unscrupulous developers were constructing houses in terraces as cheaply as possible. The Great Fire produced thousands of refugees: 13,000 homes had been destroyed in the conflagration. Most people camped in the fields on the outskirts of London while the City was rebuilt. For many the prospect of returning to the disease-ridden, fire-threatened capital was uninviting and thousands decided to make their new homes in the as yet undeveloped remaining fields of Soho which had become their temporary home. Within ten years of the Great Fire Soho was fully developed but serious fires still occurred with shocking regularity. In 1672 the Theatre Royal in Drury Lane was spectacularly burned to the ground. (It was rebuilt and burned to the ground again in 1808!)

Up until this time fire-fighting had always been something of a haphazard affair. Parties of volunteers, led more often than not by the hapless owner of the burning property, would form bucket chains in an attempt, for the most part fruitless, to quench the flames. Fire-fighting equipment was primitive and what there was usually arrived too late to prevent disaster. The Great Fire caused colossal losses to the commercial insurance companies of London and in 1667 an enterprising businessman named Dr Barbon hit on the idea of setting up an insurance company aimed at residential properties and with particular emphasis on fire cover. He established a company 'for insuring houses and buildings', and carried it on as a private venture until 1680 when he persuaded several friends to join him to form an association which was known as 'The Fire Office' – the first fire insurance office in the world. In 1688 the Fire Office was granted a charter by James II under which it became incorporated and in 1705 it changed its name to the Phoenix Office (having earlier taken for its badge the phoenix rising from the flames).

Others soon caught on to the idea: on 12 November 1696, at Tom's Coffee House in St Martin's Lane, a new insurance company had been started, calling itself 'The Contributors for Insuring Houses, Chambers or Rooms from Loss by Fire by Amicable Contribution'. It rapidly became known as the 'Hand-in-hand'. However in the course of the next three years a number of fires broke out in London which caused terrible losses, principally because there were no organized, professional fire-fighters who could be summoned to assist. Thus in 1699 the Hand-in-hand group decided to form their own fire brigade to try and cut their losses. Other companies followed hot on the trail and within ten years several groups of insurance-paid firemen were working in the City and suburbs of Westminster and other parishes.

The firemen were recruited almost exclusively from Thames watermen. Among other things, they were rewarded with immunity documents from press gangs, who were particularly fond of Thames watermen: they made fine sailors. Thames watermen had the advantage of a close knowledge of London and had a reputation of being more skilful in dealing with fires than any other class of workmen. The newly formed Westminster Fire Office began recruiting them in September 1717 when Mr Edward Wink was appointed foreman with a team of eleven 'hearty and lusty' men. The various emerging fire-fighting teams were all dressed in the most splendid liveries, proudly displaying the badge of their respective insurance companies. Some of the uniforms were outrageous and completely impractical for fire-fighting, but no more absurd, perhaps, than the uniforms their military colleagues wore for combat. Keen not to be out-dressed at a fire, one company wore bright yellow caps and coats lined in pink silk, with gold cuffs, white stockings and gold garters!

Apart from regular outbreaks of fire caused by careless disposal of ashes, over-filled fireplaces that set the chimneys on fire, candles falling over and setting fire to furnishings, bed-pans left too long between linen sheets, or whatever, there were also several instances of public rioting that ended in both tears and flames. On 2 June 1780, the day that the Whig Suffrage Bill – a very unpopular measure with the people – was going through Parliament, a crowd of 60,000 marched in protest on the Palace of Westminster. That night the mob descended on Soho, where they ransacked the Roman Catholic chapels belonging to the foreign legations; the Bavarian Ambassador's chapel in Warwick Street and the chapel in Glasshouse Street belonging to the Portuguese Ambassador were set on fire and almost destroyed.

The riots continued for several days and on 5 June 'a large mob of riotous persons' descended on Savile House in Leicester Fields. According to the *London Evening Post* the mob broke into the magnificent house, 'gutted the best part of the furniture, which they piled up in the street, and set fire to.'[1] The fires were quelled by the local parish pumps as well as the various insurance engines that arrived from all directions, eager to join in the drama and earn themselves a cash reward and plenty of free beer.

Prostitutes have been synonymous with Soho since its first foundations were laid and proved early on to be another possible source of fire. One of Soho's earliest residents was 'a lewd woman' named Anna Clerke who in

[1] Judith Summers *Soho* (Bloomsbury 1989)

The burning of the second Royal Exchange in 1838.

1641 was bound over to keep the peace after threatening to set fire to houses in Soho. During the seventeen and eighteenth centuries, prostitutes did a roaring trade in and around Soho. Each year a comprehensive A to Z of those in the trade was published under the title of *The New Atlantis* and was sold under the piazza of Covent Garden. Within its copious pages the reader would find everything he needed to know: 'the list, which is very numerous, points out their places of abode, and gives . . . the several qualifications for which they are remarkable.'[1]

There are no surviving records to show when the first insurance firemen were stationed in and around Soho, but in 1707 Queen Anne ordered that all church parishes should organize themselves with parish pumps to be manned by the warden and other parish volunteers. St Anne's in Soho, St Martin-in-the-Fields and St Giles-in-the-Fields all had parish manual pumps operated by local volunteers.

The first recorded fire 'ingeon' in Soho was built in 1720 for the enormous sum of forty-six pounds by Mr John Gray and was kept in a watch-house near Hanover Square, which was under construction at the time. A full-time foreman was appointed as engine-keeper: his principal task was to keep the engine in good order and oil the leather hoses regularly. In 1722 the engine was moved to a shed in the yard of Trinity Chapel, Conduit Street. The second engine for the area arrived in May 1730: it was much bigger and heralded as the greatest fire-extinguishing wonder in London. Unfortunately, it was too heavy for the would-be fire-fighters to pull by hand and the very first time the engine was called to a fire in Cornhill, money had to be paid out to hire a horse to pull it. The smaller engine was moved to Hungerford Market (on the site of the present Charing Cross Station) to provide more fire cover for the area. In 1733 a disastrous fire destroyed Berkeley House in Piccadilly, home of the Duke of Devonshire, and both engines were brought to work along with various insurance brigades and the parish pump from St Anne's, Soho. (The present Soho Fire Station is built on the site of the Earl of Devonshire's house, which was also destroyed by fire.)

The Sun Insurance office had a fire engine based in Swallow Street by 1789 and another at Holborn. By the end of the eighteenth century, when a cry of 'Fire' rang out in London, not one but often half a dozen different brigades would turn up. At the outset only those firemen from the company with whom the unfortunate customer was insured would fight the fire (membership of a policy being indicated by a fire-mark fixed in a prominent place on the outside of the building), but with time all firemen present would set to work and attempt to extinguish the fire, despite the relative ineffectiveness of their equipment. Meanwhile, more and more insurance brigades were formed, resulting in absolute chaos. When a fire

[1] Judith Summers *Soho* (Bloomsbury, 1989)

broke out in Soho, the parish pump would arrive, accompanied by numerous rival gangs of insurance firemen all dressed in their preposterous brightly coloured liveries. Rewards were issued to the first crews to arrive and rivalry became intense.

A large fire broke out in the Strand in 1763. One insurance company fireman, a Mr Harrison, was attacked not only by a rival fireman but then by a 'Mr Greenfield, Ensign Third Regiment of Footguards, who tore his clothes and attempted to imprison him in the Black Hole in the Savoy Barracks'.[1]

The building of the Pantheon, a vast suite of rooms between Oxford Street and Great Marlborough Street, began in the summer of 1769, and took two and a half years to complete. Its architect was James Wyatt, who lived in Newport Street. It was opened to great acclaim. Horace Walpole wrote, 'It amazed me myself. Imagine Balbec in all its glory! . . . The ceilings, even of the passages, are of the most beautiful stuccos in the best taste of grotesque. The ceilings of the ball-rooms and the panels painted like Raphael's *loggias* in the Vatican.'[2] Twenty years later it was burned to the ground. One freezing winter's night in 1792, Mr Harry Angelo, 'well known about these parts', reported with great excitement how a friend of his, an inhabitant of Great Marlborough Street, had witnessed the great fire at the Pantheon: 'I was awoke by the shrieks of females, and the appalling accompaniment of watchmen's rattles. I threw open the window, and heard the cry of fire. The watchmen and patrol were thundering at all the neighbour's doors, and people were rushing to their windows, not knowing where the calamity was seated. Mr and Mrs Siddons, who resided opposite, had, en chemise, thrown up the sashes of their bedroom, on the second floor, and called to us that the Pantheon was in flames.'

The following morning the scene was visited by many notables and thousands of curious onlookers. Amongst them was the artist J.M.W. Turner, who spent several hours studying the scene. He returned to his studio to paint a view showing the icicle-covered ravaged building surrounded by various figures, including firemen busily cleaning their equipment before leaving the devastated scene.

Theatres, too, had a particular habit of burning down and two years after the Pantheon was burned Astley's Amphitheatre was badly damaged by fire. The following year fire broke out at St Paul's Church, Covent Garden. On 17 September 1795, plumbers were carrying out work on the bell turret when, at their midday break, they left a fire unguarded which started a terrible conflagration:

All attempts to check the flames were in vain: every effort was therefore directed to the neighbouring houses and buildings which were with

[1] F. H. W. Sheppard (ed.) *Survey of London* (Athlone Press, 1966)
[2] Judith Summers *Soho* (Bloomsbury, 1989)

difficulty preserved from taking fire, so intense was the heat from the church, which was wrapt in an immense pyramid of flame rising thrice the height of the building: the heat was felt to the end of Russell Street and was scarcely to be supported within fifty yards of it. The communion plate alone was saved, but everything else belonging to the building, including the valuable and celebrated organ, the clock and other things were all devoured by the unconquerable fury of the destructive element.[1]

'Zealous and methodical they may have been, but as London began to grow rapidly with the advent of the industrial revolution, the usefulness of the individual company brigades began to be questioned. Their competitiveness had now become notorious. It was no longer simply a matter of galloping over-wildly through the streets, but had degenerated into hand-to-hand fighting and even slashing each others' hoses, while the property they were supposed to be protecting burned to the ground. By the 1830s the situation had deteriorated greatly and Mr Charles Bell Ford, manager of the Sun Fire office, proposed that ten of the largest companies should get together and form a combined force. After months of debate a scheme was finally agreed and the Alliance, Atlas, Union, Globe, Imperial, Protector, Royal Exchange, London Assurance, Sun and Westminster came together on 1 January 1833 to form the London Fire Engine Establishment.'[2]

A new era of gentlemen fire-fighters was about to begin . . .

[1] The *London Chronicle* 1795
[2] Sally Holloway *London's Noble Fire Brigades 1833–1904* (Cassell, 1973)

2 The Age of Gentlemen Fire-Fighters

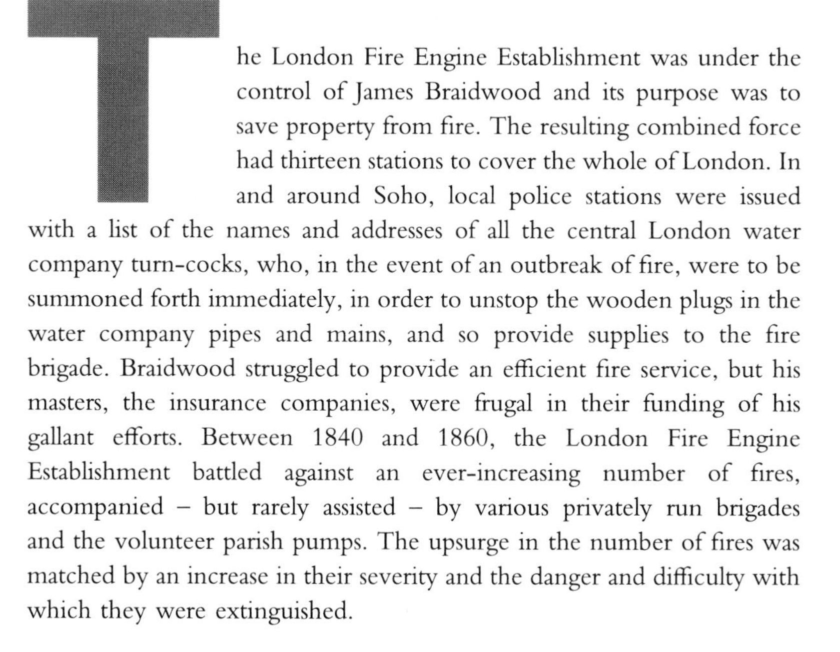

The London Fire Engine Establishment was under the control of James Braidwood and its purpose was to save property from fire. The resulting combined force had thirteen stations to cover the whole of London. In and around Soho, local police stations were issued with a list of the names and addresses of all the central London water company turn-cocks, who, in the event of an outbreak of fire, were to be summoned forth immediately, in order to unstop the wooden plugs in the water company pipes and mains, and so provide supplies to the fire brigade. Braidwood struggled to provide an efficient fire service, but his masters, the insurance companies, were frugal in their funding of his gallant efforts. Between 1840 and 1860, the London Fire Engine Establishment battled against an ever-increasing number of fires, accompanied – but rarely assisted – by various privately run brigades and the volunteer parish pumps. The upsurge in the number of fires was matched by an increase in their severity and the danger and difficulty with which they were extinguished.

One evening the painter John Everett Millais was in transit with his brother when they noticed a bright red glow in the sky. 'Accordingly we told the cabby to drive in that direction,' his brother wrote after the incident, 'and a fire engine dashing by at that moment increased our excitement. The fire was in Tottenham Court Road near to Meux's brewery, and we were in time to see the whole terrible show. On gazing upwards we noticed two firemen plying the hose as they stood on a rafter – themselves two black silhouettes against the mass of heaving flame – and I shall never forget the shout of horror that rent the air when the roof suddenly collapsed, carrying with it the rafter and the two brave men. We went home much impressed with what we had seen and my brother said, "Soldiers and sailors have been praised on canvas a thousand times. My

next picture shall be of a fireman."' His subsequent painting of this new-found urban hero was entitled *The Rescue* and was later said to be one of the artist's favourite works.

The Rescue, 1855, by John Everett Millais (1829–96), inspired by witnessing a fire in Tottenham Court Road. The fireman wears the functional uniform designed by Chief Officer James Braidwood, in sharp contrast to the outrageous costumes sported by the Insurance Fire Brigades. Brass fire helmets were introduced by Braidwood's successor, Captain Sir Eyre Massey Shaw, to replace the leather-clad helmet shown in the painting.

On 22 June 1861 a great fire took hold at Tooley Street. It was the largest London had seen since 1666 and marked a turning point in the history of London's firemen. Chief Officer Braidwood was killed, and the losses were so great it seemed that the insurance companies might collapse, and with them the entire Fire Engine Establishment. The disaster forced all concerned to re-examine the structure of fire-fighting in London. Eyre Massey Shaw was appointed to succeed Braidwood.

After years of bitter recriminations over budgets, the London Fire Engine Establishment became the new Metropolitan Fire Brigade on 1 January 1866. The old Establishment stations were taken over and another twenty-five stations constructed throughout the Metropolis. Captain Shaw was a worthy successor to Braidwood: he never shirked from first-hand fire-fighting and at the Kesterton's Carriage Factory fire in Long Acre he was thrown from a ladder, injuring his back; he took care to avoid only the most life-threatening situations.

Emphasis amongst firemen had until this time always been placed upon the protection of property, as this was the primary concern of the insurance companies which had provided and motivated the service. However, many Londoners were becoming increasingly alarmed at the growing number of lives being lost in fires, and no more so than in crowded residential areas like Soho. So it came about that in 1836, paid for by voluntary support, the Royal Society for the Protection of Life from Fire was formed. This splendid band of men, popularly known as the Escape Society, established eighty-five escape stations at strategic points throughout London, several of them were in Soho.

The organization was quite independent of the fire brigade and was set up with the sole purpose of providing wheeled escape ladders all over London to be used at night to rescue people from fires. Soho had fire escape stations at Golden Square, Oxford Street, Charing Cross, Long Acre, St James's Church Piccadilly, Tottenham Court Road and Regent Street. The escape conductor slept in a small wooden hut. On being called to a fire he would whirl his wooden rattle to summon passers-by to help him push the great heavy ladder to the fire. The ladder was made of wood, some thirty-five feet long and mounted on a spring carriage with travelling wheels. A folding ladder of twenty feet was jointed to the main ladder about ten feet from the top and if more length was needed it could be swung into position by means of ropes attached to iron levers. Some of the ladders in the West End extended to eighty feet in length and weighed nearly a ton.

For the most part these gallant escape watchmen provided an invaluable service and carried out some spectacular solo rescues. In thirteen years seventy people were saved from certain death in fires on Soho's ground alone. Eventually, in 1867, the Escape Society was taken over by the Metropolitan Fire Brigade and street escape stations became attached to full-time fire stations. Early in the twentieth century the street stations were withdrawn completely from service, to be replaced by even more fire stations built to serve the Metropolis.

All this time Soho and its environs were changing. In the early years of the nineteenth century John Nash had designed a new road link between Marylebone Park and Carlton House, called Regent Street, designed principally to mark the divide between 'all the streets occupied by the higher classes' within Mayfair, and the riff-raff of Soho village. The construction of Regent Street and of the circus at the place where Piccadilly met it provided Soho with a distinct boundary to the west.

By the late 1880s Soho was a slum filled with overcrowded, decaying tenements. Shaftesbury Avenue was constructed during 1886–87. Numbers 128–132 Shaftesbury Avenue were leased by the Metropolitan Board of Works to the London Salvage Corps to use as their western station. The London Salvage Corps had continued to work quite separately from the Metropolitan and later London Fire Brigades, funded by the City insurance companies, their purpose being to salvage the contents of burning buildings and contain where possible the damage caused to furnishings and stored properties by the water used to extinguish the fires. (The London Salvage Corps continued to operate their business until they were disbanded in 1984. Salvage duties are now carried out by the London Fire Brigade, who possess a number of salvage tenders, known as Damage Control Units, which are automatically dispatched to all fires requiring four pumps or more; their task is to prevent as far as possible damage to property caused both by the fire and smoke and by the water used to extinguish the fire.)

New firemen had to be recruited, but the number of Thames watermen available was limited, partly because several bridges were being constructed across the river. Braidwood turned, therefore, to the highly disciplined and immensely strong sailors of the Royal Navy. London firemen were exclusively ex-Royal Navy seamen until the late 1880s, and as late as the mid 1930s a strong preference was given to ex-seamen wishing to join the London Fire Brigade.

Being all ex-Royal Navy, firemen always loved spinning a good yarn and stories of the adventures of animals associated with fire stations and firemen were always very popular. The horses were remarkable for their extraordinary intelligence, their good natures and their stunning strength.

Usually they were greys, always selected in pairs by size, so that they could run side by side comfortably, and invariably purchased from the stables of Mr Thomas Tilling in Peckham. Many pairs became legendary, including Betty and Bruce at Chandos Street Fire Station. The steam fire engine at Chandos Street was named 'Fire Queen' and weighed five tons. Most stations also had pets, including cats, rabbits, pigs and parrots; many had dogs. At least ten different fire stations had a dog named 'Chance', which leads to some confusion and a plethora of conflicting stories about where and when Chance carried out his various extraordinary feats of bravery and displays of intelligence at fires. Chandos Street Fire Station certainly had a dog called Chance[1] famed for rescuing a child who had fallen into the Thames whilst watching a fire on the Embankment; escaping falling bricks and masonry at least a dozen different major fires; fetching Mr Youatt, veterinary surgeon, when one of the horses, Betty, fell ill; and insisting a fireman follow him into a burning building in Covent Garden by scratching and jumping at a locked door, behind which they found the unconscious form of a beautiful young woman overcome by smoke. The last sensation was witnessed by the Prince of Wales in person! After the fire the Prince of Wales ordered a barrel of beer to be delivered to the station, where he dropped in to swop stories of the evening's adventure with the men. When he arrived all the men were fussing around the little brown dog who had become such a *cause célèbre* at the fire. The prince stared down on the amazing creature and declared, 'I think you should call him Chance, after the dog who led Dick Tozer to the body of Braidwood when he was killed at the Tooley Street fire. He appeared from nowhere by chance, just like this one.' And so the legend continued.

The fire station was at 44 Chandos Street, running into Bedfordbury. At the back of the station were the stables, which backed on to an alley between it and the stage door of the Coliseum.

The future King Edward VII had been a fire-fighting enthusiast ever since he had witnessed his first fire, aged eleven, when the Prince of Wales Tower had burned at Windsor Castle in 1853. He cut a dashing figure as Fireman No 116, his uniform hung ever ready at Chandos Street. Tunic, boots, blue heavy cloth trousers, belt, axe and splendid gleaming silver helmet. (The firemen wore brass helmets.) Several of his close friends, including the Duke of Sutherland, the Duke of Marlborough, Lord Arthur Somerset and the Earl of Caithness, were all great amateur fire-fighters. Many evenings he would spend several hours at Chandos Street, playing cards or a game of billiards with the firemen, hoping that a fire call would come in. If the bells went down he would join them on their often hair-raising high-speed gallops to the scene of the fire, shouting at the top of his voice, helping to clear the way, 'Hi Hi Hi . . . Hi Ya Hi . . . Hi Hi . . .'

[1] Ernest Dudley wrote a book of his adventures, *Chance and the Fire Horses* (Harvill Press, 1972)

(It is probably from this tradition that the London Fire Brigade still refers to an emergency call as a 'shout'. Alternatively, it may be because of the way people shouted for help when they arrived to raise the alarm at a station, or even, some say, from people shouting into the running call telephone box on the outside of stations in their eagerness to raise the alarm.) If the prince was not at Chandos Street, but was in residence at his London palace, Clarence House, a special brake would quickly be dispatched to pick him up, convey him to Chandos Street, where he would rig, and then take him on to the fire.

Theatre fires continued to occur with alarming regularity. Captain Shaw was so concerned with the matter that in 1876 he published a book entitled *Fires in Theatres*. The book was re-published in 1889, by which time the catalogue of theatre fires on Soho's ground alone was mind-numbing. Astley's Amphitheatre was destroyed by fire in 1794, 1803, 1830 and 1841. Other theatre fires included the Circus (1805), Covent Garden (1808), the Theatre Royal, Drury Lane (1809), the English Opera House (1830), the Argyle Rooms (1830), the Garrick (1846), the Olympic (1849), Covent Garden, this time razed to the ground (1856), the Standard (1866), Her Majesty's (1867), the Alhambra (1882), the Ragland Music Hall (1883), Covent Garden – the new one – (1884), the Drury Lane (1885), Her Majesty's – again – (1885) and the Grand Theatre (1887).

On the night of 7 December 1882 the performance of another grand opera at the Alhambra, Leicester Square, was drawing to a close. At 2330 the audience had drifted away and the manager, Mr Holland, had left for home. All the duty theatre firemen had also gone home except one, Fireman Bevan, who, on making his hourly rounds of the theatre, discovered a serious fire already blazing in the balcony stalls. A few minutes before 0100, a detective named Bowden, returning off-duty to his police station nearby, saw smoke pouring out of the Alhambra dress-circle windows. He hailed a cab. 'Quick, to Chandos Street Fire Station,' where he raised the alarm. By the time Fire Queen arrived, huge volumes of smoke and columns of flame were sweeping up from the burning building. It was like daylight. Fire Queen had been quickly followed by fire engines from Great Marlborough Street, King Street, Wells Street, Golden Square, Chelsea, Kensington, Paddington, Baker Street and then Southwark, the brigade headquarters, whose jets now played on the building from all sides, but still the flames belched forth, still the smoke clouded over the night sky, filled with innumerable falling sparks. Captain Shaw, now in charge of the fire, sent requests for more appliances – urgent messages were often delivered by coach drivers riding one of their grey sixteen-hands horses at full gallop and bareback to the nearest fire station – and soon steamers arrived from Mile End, Brixton and Waterloo Road. Simultaneously, Shaw dispatched Superintendent Hamlyn to fetch the prince from Clarence House as he knew he would not want to miss such a good job.

As the fire spread in intensity, crowds of thousands of people flocked to watch the show. The flames could be seen from all over London and lit up the Houses of Parliament nearly a mile away. Shaw stood in the thick of the action, calmly directing his men, encouraging the sweating engineers to get more steam up, pump more water, feed more hoses. Every now and again he would speak to one of the water turn-cocks to ensure that the pressure in the area was sufficient for the numerous engines now pumping. Known as the 'Fire King' by the people of London, and as the 'Skipper' by his men, he stood resplendent in his glistening silver helmet, distinctive goatee beard, piercing blue eyes and hands on hips. The crowds were increasing in size by the moment. Shaw discussed the problem with a police officer just arrived on the scene, and agreed to assist by moving the crowds back as far as possible. The *Fireman* magazine waxed lyrical:

The materials of the interior were well calculated to yield readily to the flames, and once they got hold of the roof the sharp wind which was blowing rapidly fanned them into a glowing furnace. A strong force of police was soon in Leicester Square, keeping back the public, who crowded the thoroughfares leading to the place, from possible danger, and allowing the firemen room for their operations. Within half an hour the roof of the theatre yielded to the flames, which then burst up high into the air, followed by myriads of brilliant sparks, and accompanied by volumes of smoke. The square was lighted up with a noonday brilliancy, and a ruddy glow was cast over the whole of the densely populated district of Soho. The greatest excitement prevailed, the locality being one of the most densely populated in the metropolis, and the streets very narrow. The Alhambra itself is flanked on each side by dwelling-houses, hotels and schools, while behind it, extending from Cranbourn Street down St Martin's Lane, are huge blocks of property used as dwelling-houses, and many of them let out in tenements. At one time it seemed as if the whole of the east side of Leicester Square would be destroyed!

By the time the prince arrived, a desperate battle was engaged: many of the surrounding buildings were in great danger of being consumed. Various crews clambered over nearby roof-tops to get covering jets to work. The prince, always with a sharp eye for detail, spotted them and, eager to join in, followed a line up on to a roof. Superintendent Hamlyn accompanied him, as much to try and protect him from taking too many risks as to assist in the fire-fighting operations. At the front Captain Shaw glanced around

A brave fireman rescues a young girl from the clutches of death by fire.

anxiously for the prince. The fire was spreading in all directions and cracks were beginning to appear at various points across the front of the theatre. His own predecessor James Braidwood had been killed by a falling wall and he determined that neither he, nor any of his men, nor certainly the Prince of Wales would meet the same end. He turned to one of his officers. 'Go and find His Royal Highness, and ask him to rejoin me on this side of the square.'

The officer scrambled away into the chaos in search of the prince. Shaw had good reason to be concerned: the prince and Hamlyn had made their way into a narrow alley at the side of the theatre. High above them a great wave of flame washed over the roof-tops, sending showers of burning embers on to them. Prince Edward recognized two firemen from Chandos Street and greeted them, smiling enthusiastically despite the inhospitable circumstances. 'Hello, men, going well, eh?' he said, as he grasped the leather hose, slippery from the torrents of water washing around their feet. Slowly they edged forwards, working the jet close to a doorway from which flame leapt, scorching the building on the opposite side of the courtyard at the end of the alley. One of the firemen glanced back over his shoulder: their path of escape was cut off by flames which, previously subdued, now re-emerged with more vigour than before.

Seeing the seriousness of the situation, Hamlyn ordered an immediate withdrawal. With some difficulty, and not without much grunting and Naval language, the jet was turned around. Swirling smoke, carrying flying splinters of charred wood and still glowing sparks, spat down on them. Flames spouted from the windows at every floor, glaring fiercely on the spectators, who assembled rapidly from all quarters, as if defying them all and daring the firemen to do their worst. The helmets of the men gleamed everywhere amid the smoke, as they searched for weak points, turning the enemy's flanks, taking him in the side or at the rear.

Suddenly, high above them, something snapped, sending a great sizzling, spitting shower of burning hot bricks, masonry and large wooden beams down into the alleyway. Just feet in front of the prince, several men disappeared under the deadly rainfall. Without a moment's hesitation, the prince assisted in the rescue of the trapped men, burning and lacerating his hands in the process. Assistant Officer George Ashford suffered a broken back and severe internal injuries. He was removed by cab to Charing Cross Hospital, but died the following day in terrible pain. Five others, including Hamlyn, were injured.

After some time, the main roof fell in with a great crash, sending a spectacular volcano-like torrent of fire high into the air, though the two towers flanking the building stayed intact. By now still more appliances had arrived, from Castle Street, Bear Street and Green Street.

It wasn't until half-past three that morning that the firemen eventually gained mastery over the fire. The walls of the building and the minarets miraculously survived the destruction. 'Even when the fire had nearly burned itself out,' the *Era* reported in lyrical vein, 'and when Leicester Square and Soho had again lapsed into comparative darkness, the towers were blazing still, licked by circling wreaths of flame, and shone out like beacons from the smoky obscurity above.' (The Alhambra was rebuilt behind its original facade and later became one of London's most famous music halls. It was knocked down in 1936 to be replaced by the Odeon Cinema.)

The following poem by R. J. Dawson was published in the January 1883 edition of the *Fireman* magazine:

Ashford: The Fireman.

Ho! Citizens of London,
Ho! Champions of renown,
Oft need you weep for valiant men,
In harness stricken down.
Give kindly thought to those who fought
That great Alhambra fire;
Poor Ashford dead, has laurels won,
To which Kings would aspire.
At duty's call, with manly might
And courage undismayed,
Like Berg and many another mate,
They on the ruins played.
Up ladders high, that touched the sky,
On roofs too frail and weak,
Such acts as these, like heroes' deeds,
Will never cease to speak;
And ages hence, when in the dust
The Prince and Hero lie,
Such noble deeds of duty done
Shall live! While others die.

'Hurrah!' the crowd is shouting.
'An engine!' 'Clear the road!'
Here, flying round the corner,
Comes a steamer with its load!
The hose is out and ready,
There's water in the Dam,

The Suction's fixed, the pipes are laid,
The Branch Pipe has its man,
When the whistle sounds for starting
Then forth the water flies.
And it rushes up, and up;
Till it falls as from the skies.
But with tons of water falling
Upon the burning heap,
The fire roars like a furnace,
The flames still upward leap.
Hark! What was that? A shout! A crash!
And a scream above the din,
The roof! The roof has given way,
The roof has fallen in!

Now will the Firemen cease the fight?
Or dauntless courage prove?
Yes, still they face the dreadful roar
Nor backwards do they move!
Tho' a whirlwind rushes round them,
They at their post remain,
Mid the hissing, mid the roaring,
Mid the hurricane of flame!
Near tottering walls and falling beams,
'Neath sacks of chimneys hurled;
They stand like a noble army,
With banners all unfurled,
They execute their 'orders'
From the 'Skipper' they revere;
Full well they know, to him each man,
Is as a comrade dear.
For is he not a worker too,
In dangers well inured?
Yes; oft their Captain risks his life,
That theirs may be secure.

But alas! for earnest efforts,
Alas! for fervent will;
These fires so oft recurring,
Our firemen wound and kill.
And often some must say farewell
To engines and escapes,

Farewell to hose and coupling screws
And farewell to their mates.
From duty, none would shirk it,
From danger, none would fly –
But like poor Ashford, or the rest,
Be wounded, crushed, or die.
Then honour to our Fire Brigade,
Whose men possess such zeal,
Proud heroes of their native land
Their grief all hearts must feel.
And honour to our gracious Prince
Who nobly leads the van
Of public grief and sympathy,
For Ashford, brave fireman!

Captain Shaw drew attention to the dangers which were always present by the very nature of theatres. Apart from the enormously inflammable curtains and scenery, there were footlights with naked flames from either candles or, later, gaslight; there was carpentry with gluing over naked flames; there were paint shops, wardrobes and stores of highly inflammable material on the premises. Added to these were the almost unbelievably dangerous productions which managers delighted to put on – often as a token of praise for the noble men of the fire brigade!

Typical of these performances, in 1884, a year when forty-one theatres across the world burned down, killing 1,200 people, was a production at the Pavilion. Under the heading 'Fire-fighting reaches the Theatre', the magazine the *Metropolitan* reported:

Mr Fred Abrahams has produced at the Pavilion *The Streets of London*, a performance which has not been excelled in any of the West-end theatres. Viewed from the auditorium, the whole stage appears to be in flames from top to bottom and the destruction of the theatre itself seems to be imminent. There is, in reality, however, very little chance of this taking place, less, in fact, than the danger caused by wads discharged from guns in some other performances.

The *modus operandi* is as follows: the house to be destroyed is placed towards the front of the stage and perforated gas pipes are attached to the framework to increase the flames. At a convenient distance behind the scene, an iron frame is set up and covered with loosely bound tow, saturated with a light spirit such as naphthalide; a gallery upon which are pans containing coloured fire runs across the upper part of the doomed house. A 'sycopodium' pot is used to kindle the fire. The pot is made in the shape of a large pepper box and contains a sponge (saturated with spirit) attached to a wire. This fire pot is jerked about at different points of the stage and a very good representation of an outbreak of fire is produced. The tow on the iron screen is now ignited and in a few minutes the stage presents the appearance of a building which has succumbed to the fiery elements! The windows fall out (being hinged on iron frames), the gas brackets flare and the coloured fire completes the effect. It is at this juncture that one of the Messrs Merryweather and Sons 'London Brigade' Steam Fire Engines, with a full complement of firemen, dashes upon the scene, drawn by a couple of horses, with steam at full pressure and whistle blowing. Two lines of hose are run out, the firemen attack the flames, which are rapidly extinguished, and the curtain drops. In order to obviate all chances of danger, firemen are stationed on the wings throughout the performance with hoses attached to the high pressure fire-main.

Another production which must have struck fear into the soul of Captain Shaw took place at the Paragon Theatre of Varieties. It included a scene which, if anything, was more terrifyingly real than Mr Abrahams' production at the Pavilion. According to a critic's review at the time:

On the alarm of Fire, six firemen good and true, who have served in the London Brigade with Mr Howells of Chandos Street Station, headed by the popular and well-known Mr Godfrey, run the engine with a spanking pair of horses in a few seconds at galloping speed to the fire. Scene 2 finds the audience at a house well alight: the steamer comes on the ground at full gallop; with horses of which Captain Shaw might have been proud and within half a minute the hoses are run out and a jet of water is thrown into the building by an upper floor window (not stage water but solid, real water from a Metropolitan Fire Brigade jet). A first floor ladder and a fire escape are as quickly got into action. Two persons are rescued and a hearty cheer of the English type rents the air of the Paragon.[1]

Captain Shaw begged for legislation to be tightened to prevent further catastrophes, but his complaints fell for the most part on deaf ears, and at that time the Fire Brigade had no powers of legislation to enforce safety precautions.

By 1890 Shaftesbury Avenue was filled with theatres and restaurants providing the daring with a great choice of foreign food. The adjoining streets were filled with prostitutes and it was still a notorious place. In 1893 a fountain was constructed in the centre of Piccadilly Circus to commemorate the philanthropic Seventh Earl of Shaftesbury. The

[1] Sally Holloway *London's Noble Fire Brigades 1833–1904* (Cassell, 1973)

aluminium statue at its top was the Angel of Christian Charity, but so many prostitutes used to gather around its base that it was soon nicknamed Eros after the god of love, and the name has stuck.[1]

At the turn of the century the London fireman was established as a rough but romantic, heroic figure: immortalized in paintings, advertising posters, poems, theatrical productions and numerous adventure stories. R. M. Ballantyne, a prolific Victorian novelist, described his experience of riding to a fire with the men of Great Marlborough Street Fire Station:

If you have never seen a London fire engine go to a fire, you have no conception of what it is. Even if you have seen it but have not gone with it, still you have no idea of what it is! To those who mount an engine for the first time and drive through the crowded thoroughfares of London at a wild, tearing gallop, it is probably the most exciting drive conceivable. It beats steeple-chasing. It feels like driving to destruction, so wild and so reckless is it. And yet it is not reckless in the strict sense of that word; for there is a stern need-be in the case. Every moment is of the utmost importance in the progress of a fire. Fire smoulders and creeps at first, it may be, but when it has got the mastery and burst into flames, it flashes to its work and completes it quickly.

At such a time, one moment of time lost may involve thousands of pounds – aye and many human lives!

This is well known to those whose profession it is to fight the flames. Hence the union of the apparent mad desperation with cool, quiet self-possession in their proceedings . . . At times, like the roaring locomotive crashing through a tunnel or past a station, their course is a tumultuous rush and a storm of shouting and gesticulation . . .

So was it on the present occasion. The firing in the steam engine was ready laid and the water kept nearly at the boiling point by means of a jet of gas. He had scarcely applied a light to the fire and turned off the gas and they were off. Crack, went the whip, fire flew from the paving stones, fire poured from the furnace, the spirited steeds tore round the corner into Regent Street, and off they went to the fire, in the dark winter morning, like a monster rocket or a vision of Roman gladiators whirled away by a red fiery dragon. The heroes who dare anything and stick at nothing. Had the fire been distant, they would have had to commence their gallop somewhat leisurely for fear of breaking down the horses, but it was not far off – so they dashed round the corner of Great Marlborough Street at a brisk trot and swept out into Oxford Street. Here, they broke into a gallop and here the noise of their progress began, for the great thoroughfare was crowded with vehicles and pedestrians, many of whom were returning from the theatres and music halls and other places of entertainment.

Charles E. Stewart's breathtaking painting of a Metropolitan steamer galloping full tilt to a shout.

[1] In the early 1980s, as part of the redevelopment of Piccadilly, the statue of Eros was taken down and cleaned. However it was restored so beautifully to its original gleaming aluminium that it was decided it must be re-darkened to make it look as it had before: after much trial and error a solution was found – London Fire Brigade boot polish!

To pass through such a crowd without coming into collision with anything required not only the most dextrous driving but rendered it necessary that some men on the engine should stand up and shout, or rather, roar incessantly as they whirled along, clearing everything out of their way and narrowly escaping innumerable crashes by a mere hair breadth.

But in reality the lives of these firemen and their families were far from glamorous. They worked a continuous duty, and very rarely had time off. Pay and general conditions were appalling. The families lived together above the station. The children led isolated lives and had few friends. Sally Holloway's father, Douglas Gray, was born into such a family. His father, Peter, was based at Great Scotland Yard Fire Station from 1899 to 1902. During the day his children attended the St Martin-in-the-Fields school, and had so little money that they could not even afford boiled sweets from the local shop; just the crumbs from the bottom of the jar.

The first horse-drawn escape ladders (known as horsed escapes) were introduced to London in July 1897. The first time a horsed escape was used in a rescue was by Soho's crew on the evening of 17 October 1898:

News came that a serious fire had broken out in Oxford Street. The extensive premises of Messrs E. Tautz & Co, wholesale tailors, were discovered to be in flames and the alarm was brought to the fire stations from various sources.

The Orchard Street fire alarm rang into Manchester Square Station and resulted in the horsed escape being turned out; then another alarm rang into Great Marlborough Street Fire Station and the horsed escape had hurried from this point also. The appliance was new, and for some time the men of the brigade had cherished a laudable ambition to be the first to use the escape in what they call a life-saving job. And it was only by an untoward chance, or simple fortune of war, that the men of the Manchester Square Station, who were first on the spot, missed the coveted honour.

When they arrived on the scene, no sign of fire was visible in Oxford Street itself, and the firemen were pointed to North Row, one of the boundaries of the burning block behind. They made their way thither, searching for inmates, but were driven back by the fierce flames.

Meantime, the three persons sleeping on the premises – the foreman, Mr Henry Smith, his wife and their little son, aged six years – had been endeavouring to escape by the staircase, but had been driven back by the dense smoke filling the room, and he aroused his wife at once and took the boy in his arms. Not being able to escape by the staircase, they hurried

to the front of the large block of buildings, shutting the doors after them as they went. So it happened that they appeared at the second floor windows facing Oxford Street just as the horsed escape from Great Marlborough Street Fire Station hurried up. A scene of great excitement followed. The firemen ran the ladders from the escape to the building, and brought down all three persons in safety, but Mrs Smith unfortunately had suffered a burn on the left leg. It is probable that, but for the rapidity with which the horsed escapes arrived on the scene, the family might have suffered much more severely; for the fire was very fierce, and soon appeared in Oxford Street.

The crew at Great Scotland Yard Fire Station, c. 1900. Peter Gray sits far left, middle row. Behind them, their glorious steamer, endlessly polished, gleams in the morning sunlight.

The honour, therefore, of the first rescue by the new horsed escape rests with the Great Marlborough Street Station, though the efforts of their brave comrades of the Manchester Square Station should always be remembered in connection therewith. Commander Wells, the Chief Officer, appreciated this, for he telephoned a special message to Superintendent Smith saying, 'Please let your men understand that I thoroughly appreciate and approve their action on arrival at the fire this morning, although the honour of rescue falls by the fortune of war to the second horsed escape.'[1]

By 1900 electric bells had been fitted in stations to alert the crews when an emergency call arrived. Every flat above the station was fitted with two bells: one in the bedroom and one above the front door in the hall. All the men were ex-sailors – Peter Gray had run away from home aged eight to join the Navy – and they were all strong as oxen. His son recalled how 'they used to have competitions in the yard to see who could lift a manual pump by lying on their backs and raising it on their hands and feet – a dead weight of one and a half tons!'

In 1882 a new fire station had been built in Great Scotland Yard as part of the redevelopment of Northumberland Avenue. The men and machines of Chandos Street Fire Station moved to their splendid new home in 1883, when Chandos Street closed. The station's complement now consisted of one station officer, eleven firemen, two coachmen, two pairs of horses, one horsed escape, one steam fire engine, one manual escape, one long ladder and one hose cart. The station was given the new number of 64. The nearest station to the east was Whitefriars, Station 62, and to the north, Great Marlborough Street Fire Station, Number 72.

Thus the 'ground' covered by the present day Soho Fire Station (Number A24) was originally served by four stations: Great Marlborough Street – under the command of the B Division with its headquarters at Clerkenwell – Holborn, Whitefriars and Great Scotland Yard.

[1] F. M. Holmes *Firemen and Their Exploits* (S. W. Partridge, 1902)

INTO THE TWENTIETH CENTURY

Throughout the early years of the twentieth century the Metropolitan Fire Brigade (re-named the London Fire Brigade in 1904) became involved in the rapid mechanization of its equipment. One of the most important additions to the armoury of tools the firemen had at their disposal was the hook ladder. In 1902 a dreadful fire occurred in Queen Victoria Street during which nine people lost their lives, trapped on upper floors which could not be reached by the fifty-foot escape ladders. The hook ladder originated in 1826 in France, where they were called 'Pompier Ladders' and by the 1880s many European fire brigades used them, as well as brigades in large American cities where the ladder became a central piece of rescue equipment, giving its name to the now legendary 'Hook and Ladder Companies'. Had hook ladders been available to the men of the London Fire Brigade at the Queen Victoria Street fire, those nine lives would almost certainly have been saved. The following year hook ladders were introduced to the London Fire Brigade. The ladder was just thirteen feet in length and weighed twenty-five pounds; it was made of wood fitted with a hinged tensile steel hook with a serrated edge. The ladder enabled firemen to scale a building in order to gain entry at any floor for the purpose of effecting rescues, searching or general fire-fighting. From the first year it was introduced operationally, the hook ladder proved its worth, assisting in the rescue of hundreds of people who would otherwise have perished beyond the reach of other ladders or at the inaccessible rear of buildings where no other ladder could be pitched. Another improvement in equipment was the replacement of old leather hose with canvas hose.

The fifty-foot escape ladder was still the primary rescue ladder, but its transportation to fires, which had been so slow when pushed by hand, was now horse-driven, mounted on a specially adapted carriage. As a result the last of the street escape stations had gone by 1904. Electricity, steam and new-fangled petrol-driven appliances were designed and used experimentally, but it was quickly established that the internal combustion engine had the lead and by 1921 all the horse-drawn appliances had been replaced by petrol-driven ones. Soho Fire Station (still at 53 Great Marlborough Street) lost its horses at the beginning of 1914.

In 1906 the first turn-table ladders were introduced to the London Fire Brigade. These revolutionary metal ladders reached a height of eighty-two feet, and were fitted to horse-drawn articulated carriages, upon which the base of the ladder could turn through 360 degrees, giving the operator manoeuvrability far superior to anything that had previously been available. Soho did not receive its first set of ladders until 1937 – until that time, unlike the City, there were very few high buildings on Soho's ground.

As the mechanized appliances were introduced, so they were fitted with large bells which the officer riding in charge of the machine, always seated to the left of the driver, could give a really good hammering, to clear the way and herald their imminent arrival to any person in dire distress.

Early experiments in breathing apparatus were also started at the beginning of the century, and because Soho was a station with a reputation for picking up a lot of working fires, new equipment was often tested out on the men at Station 72 (and this is still true today).

With the outbreak of the First World War, there was great excitement because of the air-raids. Policemen from the station in Great Marlborough Street would ride up and down the Soho streets, wearing large placards around their necks, declaring 'First Warning' and then more alarmingly 'Take Cover'. Other locals acted as self-appointed wardens who were more like town criers, running down the street ringing a large hand-bell or blowing a hunting horn, while shouting warnings to all the residents. 'Take cover, take cover!' they would yell, in such a frightening manner that young children in the street would burst out crying. 'A terrible attack from the air is coming. To the shelters!' Harry Errington, later to be awarded the George Cross for his bravery as a fire-fighter during the Blitz, was only five when the raids began; he remembers them as tremendous adventures:

We would then dash as fast as we could to Oxford Circus Underground Station and down on to the platforms. It was so exciting as we would go on joy rides all along the line and back to pass the time until the raid was over. Each carriage had its own guard, who would push and pull a big lever to open and close the doors to his carriage. When we eventually returned to Oxford Circus, we would emerge into the streets and run around in great screaming gangs, collecting bits of shrapnel off the pavements and from the gutters, bantering all the while as to whose was the hottest piece. It was always way past our bed-times and that made our pleasure all the greater.

During the First World War the brigade lost many men killed in action who had been called to the colours, but after enemy raids began on London in 1915 firemen were exempted from conscription. This was rescinded for a time but after serious raids on central London in 1917 all members of the brigade serving abroad were again recalled.

There were a total of twenty-five air-raids on London. Compared with the Blitz of the Second World War the figure seems very small, but at the time the whole concept of aerial bombardment on the capital city of the still extensive British Empire seemed unthinkable. Seven of the raids were carried out by airships, and the rest by aeroplanes. The first came on 31 May 1915, the last on 20 May 1918. During that time 922 bombs were dropped of which 567 were explosive and 355 incendiary. In February 1918 a particularly concentrated raid occurred with many areas on Soho's ground being hit, including Covent Garden, the Strand, Holborn and most disastrously the printing works of Messrs Odhams in Long Acre. This large building was considered an excellent air-raid shelter and more than two hundred people had taken refuge there when suddenly a high explosive bomb made a direct hit on the central part of the building, killing thirty-one people and seriously injuring more than a hundred others.

Other deadly raids on Soho's ground included Zeppelin attacks on Piccadilly and the Aldwych, and aeroplane bombings on Broad Street, Charing Cross Road and Piccadilly, again, when the Royal Academy of Arts was hit twice. Another bomb landed on the Embankment, causing shrapnel damage to Cleopatra's Needle which is visible to this day.

After the First World War a fire brigade's union was formed which, after much campaigning, won their members the right to take one day off in ten. This consideration was greatly increased when the powers that be agreed to the formation of a two-shift system (named Red and Blue Watches) which meant for the first time ever firemen and their families were no longer compelled to live at fire stations (though as each station only had one Station Officer, he was still required to live on the premises). But the number of firemen had to be doubled to cope with the new shift system.

The war had a profound effect on the social structure of British society and the availability of manpower. As Judith Summers, in her book *Soho*, writes:

BIRCH BROS TAILORS
158
BIRCH
TAILORS
BIRCH
BROS
EXPERT
TAILORS
DAY
HAIRCUTTING & SHAVING SALOON
HARRIS
157
FINCH'S
WINES
FROM THE
WOOD
FINCH
FINCH'S
FINCHS
BIRCH
TAILORS
8/-

By 1918, the formal order of Victorian England had been swept away for good. There was a new feeling of egalitarianism about, which Soho's atmosphere suited down to the ground. With former domestic servants who had been conscripted into the army or war-work now disinclined to return to their old jobs in service, 'eating out' in restaurants became an increasingly popular pastime for the moneyed classes, who, like their servants, had been liberated from certain social conventions by the war . . . Since the Oxford Street gown stores were still very expensive, a whole new breed of shops selling cheap, ready-to-wear frocks opened in Berwick Street market to cater for the new working woman with a little money in her pocket. Soon nearly every front parlour from D'Arblay Street to Shaftesbury Avenue had been replaced by a shop window.

Opposite: A serious fire has broken out in the premises of a photographer at number 159 the Strand, adjacent to King's College, c. 1905. The firemen have just arrived. 'Slip the escape!' bellows the officer. 'Get jets set in!' All around him crews rush into action. White canvas hose snakes across the pavement as the escape crew pitch the ladder to the top floor. The 'horsed escape' wagon is visible on the right of the picture.

Thus in the space of a few years the whole nature of the fire ground changed. Virtually every house in Soho opened a shop on the ground floor, selling goods, clothes or food; dozens of other premises were converted into restaurants. In the twenties and thirties there was no fire legislation and little or no fire prevention work carried out by the fire brigade. As a result, the area became a very high fire risk. Sweat shops sprung up above and behind shops, with dozens of people working in confined spaces, the exits blocked by packing cases, boxes, rolls of material or whatever. Kitchens were created in crowded basements converted from cellars, store rooms and dingy, dark bedrooms. Almost every domestic house, already overcrowded with immigrant families from all over Europe, lost its ground floors in the interests of business and profit. Large new buildings sprang up on the ground too: Regent Street was completely rebuilt on a grand scale. Banks, corporations and other international companies constructed new monuments to their achievements and by the mid thirties the whole character of the West End, and particularly Soho village, had changed. Most of the firemen had seen service in battle: they were a hardened, highly drilled, tough group of workers who thrived in the disciplined atmosphere encouraged at the fire station.

In 1921 the London Salvage Corps moved out of their splendid building in Shaftesbury Avenue, the freehold having been purchased by the London County Council, and the fire stations at Great Scotland Yard, Whitefriars and Great Marlborough Street (Number 72) were all closed to re-emerge together as the new Soho Fire Station (re-numbered 72) in Shaftesbury Avenue. To one side of the station was the famous Avenue Bar, and to the other, the Shaftesbury Theatre. Opposite was the Palace Theatre, which took to playing musicals only because of the noise the firemen made on turning out to a shout. (Even after two-tone air-horn sirens were introduced, it was a strict Soho tradition, as a matter of

Overleaf, left: Three people are trapped on the top floor of a building in Fleet Street. The appliances from Whitefriars Fire Station have already arrived. One fireman, without tunic or helmet, has reached the top floor using a hook ladder. A colleague hauls a second hook ladder aloft while the rest of the crew feverishly wind the handles of the escape, carrying a third fireman already at the head of that ladder towards the people trapped.

Overleaf, right: About a minute later. The fireman at the top window has repositioned his hook ladder on the next-door windowsill, making room for the head of the escape to pitch to the window. One woman, her dress billowing in the wind, is already being carried down, while a man is being helped on the top of the escape. The fireman with the other hook ladder has reached the second floor, and is about to enter to carry out search and rescue operations.

HORACE MARSHALL & SON
PUBLISHERS OF
OFFICES TO LET
THE METHODIST
BY AUCTION THIS DAY
BANKRUPT STOCKS!
DAILY

OFFICES
TO LET
THE METHOD
SALE BY AUCTION THIS DAY
BANKRUPT STOCKS!
DAILY

courtesy, to turn out of the station 'on the bell', but not using the very loud, penetrating air horns.)

Soho's firemen continued to battle with the regular number of emergency calls the area threw at them. On 23 October 1924, at a time when unemployment was high and the depression was galloping, but when the British film industry was thriving, a serious fire broke out at a building on the corner of Wardour Street and Meard Street. The building was occupied by several different companies all involved in film production and contained an enormous quantity of celluloid film. On arrival, Soho's crews found that the fire had gained a great hold on the building, and thick black smoke issued from windows on all floors. Soon afterwards, a great explosion occurred, showering the street with hot masonry and shattering all the windows of the buildings opposite. So much blazing debris fell down into the street that the appliances had to be withdrawn to a safe distance, but too late – a car and a van parked in the street burst into flames! All fifty people working in the building escaped unharmed, but the building was reduced to a smoking ruin.

The following month, during the night of 25 November, a fire broke out in an annex of the British Empire Club in York Street, St James's. Soho's pump escape and pump arrived within minutes of the bells 'going down' (the fireman's expression for an emergency call arriving at the station). The men's polished brass helmets glinted in the light of the street gas-lamps, and reflected the orange flames licking from the building as they leapt into action. The lift shaft of the elegant club was ablaze from the ground floor right up to the top, as was the grand staircase which ran around the shaft. All the occupants who had retired to their rooms found their escape route cut off, and now appeared at windows on the fourth and fifth floors, both at the front and back of the building, shouting and screaming to be rescued. In four minutes no fewer than ten people were rescued using hook ladders, although sadly the housekeeper, who lived on the top floor, died before they could reach her, suffocated by smoke. Soho's firemen were dazzling in the speed and skill they showed in the rescue of so many people. All those hundreds of hours of training to use the hook ladders safely and efficiently had paid off a thousand fold.

In 1933 another serious fire broke out in Wardour Street. Fireman Reg Johnson, who had joined the brigade like his father before him, carried out a particularly hazardous rescue of a lady from the second floor. When Reg climbed into the room the lady was naked, but to his great relief managed to slip into a wispy nightdress, 'which did little to protect her modesty', before she collapsed into his arms. He reappeared at the window, grasping her firmly, and flung her over his shoulder in the traditional fireman's lift so beloved of Victorian illustrators. The entire rescue was captured on film by a newsreel cameraman, though sadly the film seems to have been lost.

In 1933 Major C.B. Morris was appointed Chief Officer of the brigade and among other innovations he was responsible for introducing 'dual-purpose' appliances. These machines could be used either as pumps or as escape carrying vehicles, which made it possible to dispense with the old single-purpose escapes. He also introduced the enclosed body for appliances, giving the crews some protection from the weather and in 1935 Soho was issued with a splendid limousine sloping-back enclosed pump. The pump escape was still open air, which meant that on a bad winter's night the crew would arrive at a shout with their hands frozen to the metal hand-rails to which they had to grip with all their might for fear of being thrown off the fast-moving 'red devil'.

The firemen were no longer all ex-Navy but the brigade retained many of the Naval traditions, including the ringing of bells on stations to indicate changes of watch and stand-easy periods, and many verbal expressions that owed their origins to the Navy. One such phrase which the officers would always shout at new fireman was, 'Remember, keep one hand for yourself and one hand for the fire brigade.'

On his first day at Soho, a very young Georgie Phillips was taken to one side by his new Guvnor. 'Listen, Phillips, when we get a shout, I want you to keep one hand for the brigade, and the other for yourself. You grip hold of the side of the machine for all you're worth, or else we'll lose you!' Suddenly the bells went down! Georgie struggled into his squeaky clean fire tunic, leather boots and helmet, did up the buttons as fast as his shaking fingers would allow, pulled on his belt and axe, and jumped on to the side of the escape. He clasped his right hand around what he thought was the handle on the side of the machine, and braced himself for the first ride of his life. Like a charioteer in *Ben Hur*, he rode out into the chaos of Shaftesbury Avenue in the 1930s, dozens of passers-by stopping to point and stare: a glorious, glittering patchwork of London life rushed past Georgie in a high speed blur as they raced to a street fire alarm that had been actuated. When they arrived Georgie glanced down at his right hand. 'It was almost numb from gripping so tight. Then to my horror I realized I hadn't even been holding on to the fire engine. My knuckles were white with gripping on to the handle of my axe! I'll never know how I wasn't thrown to my death on my very first shout!'

In 1936 the magnificent brass helmets of the London Fire Brigade, introduced by their most glamorous chief Eyre Massey Shaw, were withdrawn from service. The use of electricity had become widespread and wires exposed in fires became a great hazard to the firemen in their metal headgear. Thus a new cork helmet was issued as a replacement (and continued in use, with some small changes in design, until 1990).

Fireman J. W. Root steps forward to receive the Council's Silver Medal for Gallantry from Lord Snell, Chairman of the Council, in recognition of his extraordinary bravery on the occasion of a fire at 5 Peter Street on 29 February 1936. The firemen wear axes on their left hips and hose spanners on their right. These spanners were used to tighten round-thread hose-couplings and remained in use until after the Second World War when, under the auspices of the National Fire Service, automatic-locking standardized hose-couplings were introduced to brigades throughout the country.

The Station Officer at Soho during the 1930s was Mr Jeffreys, who still lived above the station with his family. The two watches were run day-to-day by the Sub-Officers, but Mr Jeffreys rode on the pump to every fire call, twenty-four hours a day and with very few days off. During the thirties rivalry between stations was greater than ever before. Numerous fire service drill competitions of various descriptions were a regular feature of brigade life, as were sporting competitions. The brigade had a particularly fine boxing team, which was run by Mr Jeffreys at Soho. The team were very successful, made up principally of ex-Navy and Army boxers, but the star of the team was Tony Stuart, holder of the coveted American 'Golden Gloves' award, and he was a Soho man.

The West End had dozens of automatic fire alarms and fire telephones (direct lines that connected the premises to the fire station): all the theatres and important buildings had fire telephones and there were also hundreds of street fire alarms, which were the principal cause of false alarms. (The street fire alarms were eventually withdrawn from service in the late forties, as exchange telephone public call boxes were fitted in streets, as well as an increasing number of telephones in business premises and private homes, and the 999 emergency telephone number was introduced as part of the rapidly expanding telephone service to Londoners.) As a result, even in the 1930s it was a very busy ground. The watchroom, which was still permanently manned, buzzed every morning and evening as the street alarms had to be tested regularly – that was always a popular job as the pubs around Covent Garden market opened first thing in the morning especially for all the traders.

The firemen and Sub-Officers worked a two-watch shift system: a week of days, followed by one day off, and then a week of nights. In their spare time, they all did part-time work to supplement their incomes, as the pay was dreadful. Shifting scenery at Covent Garden Opera House, grave-digging, coffin-carrying and popcorn-making at a small factory in Soho Square were all very popular part-time jobs. Some of the firemen had more specialized skills including jewellery-making and watch repairs, and one, who became known as Doc Ward, was a chiropodist in his spare time: 'He used to come to work stinking of ether, but he did our feet as well, so he was all right!'

Soho Fire Station was now established as one of the most desirable postings in the London Fire Brigade. There was nowhere else quite like it anywhere in the world, let alone London. Whenever a fire broke out on the ground, whether in a large Palladian mansion in St James's, a Corner House dance palace in Leicester Square or a crowded tenement block off Berwick Street, the fire would always demand the total resources of every fireman present to prevent catastrophe and conflagration.

But no amount of serious fire-fighting in the twenties and thirties could have prepared the men of Soho for the fires, and the human horrors, that were about to be unleashed upon them.

16 April 1936: this car burst into flames while driving along the Mall. The pump was quickly on the scene, but the car was already burned out. The two firemen are wearing the new cork helmets while their Station Officer still wears his brass one. Soon after this picture was taken, Station Officers were issued with new black cork helmets, with the comb painted white to indicate their rank.

Fire at HMV, Oxford Street. All crews have been withdrawn from the building, which is in danger of collapse. Four ground monitors pump jets of water into the blazing building, and two turn-table ladders – one from Soho – act as water towers, directing powerful jets into the upper floors. Three wheeled escape ladders stand in the street, having been withdrawn from the front of the building.

The Second World War: Soho's Burning

4

By 1937 the threat of war rumbled like distant thunder across England's green and pleasant land. Under the Air Raids Precautions Act the fire service began to recruit additional personnel to boost its resources. The Auxiliary Fire Service was established. Most of the recruits were men but there were a few women. Nine in fact.

These first ladies were welcomed into the AFS, having been selected for the skills they already had from their full-time employment. One of them was Jessica Ireland (née Underwood) who was born in 1904 in Streatham Street, where she still lives. She was a post-office-trained telephonist and switchboard operator and should have been sent to Lambeth Fire Brigade Headquarters, but she persuaded them to let her go to Soho 'as it was so close to home, and I worked in the City during the day.'

The girls were given intensive training in operating switchboards, administration, driving, map reading, running station watchrooms, and all basic aspects of fire-fighting, despite their rather cumbersome skirts. 'As long as I live I will never forget the day we persuaded Sub-Officer Barber to let us try the pole. We all lined up in the corridor, he gave us a brief instruction on how to step forward and grasp the pole and then we were off!' Jessica was first, followed very smartly by her colleagues. Unfortunately Mr Barber forgot to tell the girls to let go of the pole and stand out of the way as soon as they reached the bottom, 'so there was the most almighty pile up with legs and skirts going everywhere, here and there and all in the air!'

On another occasion the girls were given a lecture by the ever-patient Sub-Officer Barber on gases. Jessica Underwood was sitting in the back row at a desk next to the window looking out over Shaftesbury Avenue. Outside the Palace Theatre a couple of buskers were playing:

They were jolly fine musicians and I got rather distracted. I was handed this glass test-tube with the purpose of having a quick sniff so that we could recognize the different kinds of gases. This particular one smelt of geraniums, and was really quite pleasant. So I had another little sniff. Then a third, while all the time I was watching out the window. I had a fourth without thinking, and next second I was flat on my back. Talk about a gas attack: I gassed myself!

The watchroom was dominated by a large wall map of the area, a splendid round brass clock, a bank of telephones and the dials relating to automatic fire alarms. In the centre of the room a long wooden cabinet stretched from end to end, covered with glass revealing numerous typewriter-like hammers. This complex machinery was linked directly to the street fire alarms which dotted Soho's ground. The alarms were about the height of a modern parking meter displaying a round panel with a glass screen. In the event of discovering a fire the user smashed the glass and pulled the handle inside the head of the street alarm. As soon as it was pulled the message was relayed to the station watchroom where the machine would type out a number code. The watchroom operator then looked up the number of the alarm on the wall to identify the location. As soon as this was done the watchroom girl would put the bells down.

One evening in early 1939 Jessica Underwood was on duty in the watchroom when the 'running call' bells began to ring (in other words, a member of the public had come to the station to raise the alarm by opening the door on the telephone cupboard situated on the outside wall of the fire station). She opened the front door to be greeted by an agitated policeman. 'Quickly, love, there's a bad one going in Dean Street and there are people involved.' The turnout was rapid but the building was already ablaze.

'It was a right proper prostitute's place,' Jessica later recalled. 'Some poor girl and her customer were caught on the top floor. They made their way on to the roof but the flames came gushing out at them and they both threw themselves to the ground before the firemen had even arrived.'

Soho was a busy station and during the 'phoney war' period the regular 'red riders', as the old London Fire Brigade men were known (because they rode on the pre-war fire engines that retained their red livery), responded to all the usual calls to fires and numerous false alarms caused by vandals activating the street alarms. At the sub-stations training continued every day and many more men and women were recruited into the fire brigade. The girls who had been in the job since 1938 were almost 'old hands'. During the pre-war period they were even taken to large fires to witness their male counterparts' actions first hand.

As war approached the firewomen too became involved in training new girls to man the watchroom as well as do all the administrative work; they would help in the canteen too. Once the Blitz started they carried out the most unbelievably dangerous tasks on the fire ground itself – equal in every way to the dangers faced by their male colleagues: as staff car drivers they had to manoeuvre their way through the rubble of collapsed buildings, nipping down unaffected side streets but continually risking the imminent downfall of burning rubble from above. Once at the scene of a major fire they would be required to stand by their staff car ready at a moment's notice to ferry the officer in charge to another conflagration. Others rode on motorbikes as despatch riders, delivering vital messages between control rooms, fire stations and the scenes of the fires. One or two even drove fully laden petrol lorries through the streets of the West End at the very height of the Blitz, to refuel the pumps, while bits of shrapnel, burning embers, anti-aircraft shells and high explosive bombs and incendiary bombs rained down on them from above.

In order to provide the anticipated necessary fire cover for London the authorities ordered the conversion of several hundred properties into fire sub-stations, organized into groups attached to a parent regular fire station. Each sub-station had a small fleet of fire appliances: mostly London taxi-cabs painted dull grey and converted to pull 'trailer pumps', carrying the minimum of equipment: hose, axes, hydrant bars and keys and rescue lines (ropes). The requisitioned properties were chosen as they all had garage facilities either attached to them or very nearby. Soho's six stations were in a car showroom in Jermyn Street (which moved to the Canada Life Building in Charles II Street), Jackson's Garage in Rathbone Place, the Odhams Press printing works in Long Acre, schools in Peter Street and Adelaide Street, and the City Literary Institute in Stukeley Street. In addition there were forty-eight street fire alarm posts across the ground. These were similar in construction to a gun-emplacement, with sand-bags piled about ten feet high to form a small shelter, usually on the corners of large streets. Once the Blitz began the 'posts' were continually manned by a fire crew of four or five men, ready at a moment's notice to rush into the streets nearby and extinguish incendiary bombs. Each auxiliary station was staffed by a nucleus of experienced London Fire Brigade men and officers who between them were tasked with the high-speed training of their new complement.

Throughout the spring and summer of 1939 more and more recruits were welcomed into the AFS. Once trained they returned to their normal employment but were ordered to be on stand-by to report to their stations as soon as the call-up was declared. On 1 September 1939 the call came.

On that fateful day dozens of fresh-faced young men and women made their way to Soho Fire Station. It was an impressive building dominating the south side of Shaftesbury Avenue near Cambridge Circus. Above the station doors the brass lettering read 'London Fire Brigade', but underneath you could still make out the shape of the previous lettering, 'London Salvage Corps', reminding all who noticed of the station's heritage.

In addition to the firemen and women arriving at the station, there were also a selection of young, fit youths armed only with bicycles: they were the messenger boys. The minimum age was sixteen. Little did these enthusiastic, bright-eyed children know that within a year they would be dashing through the streets of Soho they knew so well in scenes second only to the Great Fire of London itself, delivering urgent despatches to and from the scenes of fire and destruction.

On the morning of Sunday 3 September the crews at Sub-Station X[1] gathered around the wireless. The voice of Prime Minister Neville Chamberlain crackled his message: 'I am speaking to you from the Cabinet Room at 10 Downing Street. This morning the British Ambassador in Berlin handed the German Government a final note stating that unless we heard from them by eleven o'clock that they were prepared at once to withdraw their troops from Poland, a state of war would exist between us. I have to tell you now that no such undertaking has been received, and that consequently this country is at war with Germany.'

For an instant the whole of London let out a spine-chilling sigh, then the air-raid sirens shrieked their terrible wail over the city. People in the streets ran for cover and panic spread through the air. In the underground watchroom of Sub-Station X, a young girl lost her head with fright and collapsed in convulsions of screaming panic. The officer in charge ordered her to be removed. 'Get me someone else from Shaftesbury Avenue.' The replacement was Violet Parsons. Fate threw her back together with a former flame, a Jewish lad called Sam Goodman, whom she nicknamed Benny, and before long they eloped together (she being from a Christian family) to be married. They recently celebrated their fiftieth wedding anniversary. Said Sam:

Soho brought us together and we've had a marvellous happy life together ever since, and we often remember the friends we made during those times: there were some truly wonderful characters at Charles II Street: Bryan Gibbens – he later became a judge; Eden, the pianist; Richard Selby – he was a Cambridge lawyer; Mr McIntevy – a typical Irish, ex-Navy, old LFB man – strong as an ox, hard as nails, heart of gold. There was also a very fine portrait painter by the name of Briggs. He actually painted my

[1] Confusingly the sub-stations were re-allocated letters in 1941 as follows: T-Peter Street, U-Charles II Street, V-Adelaide Street, W-Long Acre, X-Stukeley Street, Y-Berners Street (which replaced Rathbone Place).

portrait, but unfortunately my dear mother used it to help draw the fire in her coal stove, so that went up in smoke.

Another local man who joined the service was posted to Rathbone Place: Harry Errington was born at 47 Poland Street on 20 August 1910, and has remained a Soho resident throughout his life; he continues to live and prosper there as one of the West End's finest tailors.

Having worked already for a number of years as a tailor in and around Piccadilly and Savile Row, he walked to the fire station in Shaftesbury Avenue to volunteer as an auxiliary fireman, a few days before war was declared. On arrival he found there was no waiting on ceremony: he was interviewed, looked up and down, given a quick fitness test and signed on.

We were given our uniform on the spot, but I had such a big head they couldn't find a helmet to fit me for months. As soon as we were enrolled and kitted out, we were then sent out to our postings at one of the sub-stations. I was ordered to report to Sub-Station Z on the corner of Rathbone Place and Oxford Street. On arrival I found several converted taxi-cabs with trailer pumps parked in the street, along with a group of other firemen including some regulars who were our instructors and mentors. I asked the officer in charge, Mr Skinner, 'Where do I put my bedding, sir?' He and the others all laughed at me. 'There's nowhere to put it, lad, because there ain't no beds. We sleep in the back of the taxis!' At first I thought he was joking, but it quickly became obvious that he was not. And so it was that we slept for several weeks crunched up in the back of the taxi-cabs, while not two hundred yards away the rest of my family slept in the comfort of their lovely beds.

After a couple of weeks we moved to a school in Hanway Place whilst our appliances were found a new home in Jackson's Garage, a large building of three floors and a basement with its entrance in Rathbone Place. Customers would drive in at ground level and either park on the ground floor or else drive on into the elevator which would take cars to the second or third floor. We parked all our vehicles and equipment on the ground floor. Whenever we got a shout at night there would be this crazy dash: a quarter of a mile charge from our sleeping quarters to the garage. This was fine for the young and healthy boys, but it nearly finished off some of the older men.

During the phoney war we were always busy drilling, preparing for the worst. Despite the fact auxiliary firemen were generally held in low esteem by the general public, the atmosphere at the station was excellent: there were a number of fine chefs in our squad, and we were served excellent food in our mess. There were also a number of men from the theatre world, and we quickly organized a first-rate concert party. One particularly funny fellow named John Terry did a farcical striptease, and my star role was to push my arm through the curtain to act as a hook for him to hang his clothes on as he performed. It used to bring the house down.

Soho itself was a very happy station, notable for its feeling of goodwill, comradeship and co-operation. There were no fights or quarrels and romance blossomed more than once. 'One of the two Misses Teulon sisters married a Soho boy and Assistant Group Officer Pullen also married a Soho fireman,' Jessica Underwood recalls with romantic eyes. These plucky girls were commanded by several notable characters. One of them was Company Officer Bentley 'who was a real stickler! Whenever the bells went down she would order us to drop whatever we were doing, pull on our tunics, dash down into the watchroom, stand with our backs against the wall – so as to keep out of the way of the men – and wait and watch. The old LFB firemen used to tease us all so much about this drill, but it paid off by saving our lives the following year.'

After the initial excitement of the declaration of war, life for London's new fire-fighters became difficult. News of the fighting in Europe was bleak and many firemen were dubbed 'three-pound-a-week war dodgers'. So many began to quit that the government passed a statutory order preventing full-time fire-fighters from leaving. 'One fellow was Harry

The entire squad of Rathbone Place Sub-Station standing on the roof of Jackson's Garage. The regular LFB men, wearing peakless caps, are to the left. Harry Errington, hatless, sits in the middle. The two men he later rescued are John Hollingshead, back left with round glasses and the crossbelt of his respirator, and John Terry, back row eight from the right.

Dolly who went off to the Navy for a time, but he was sunk so many times that eventually he came back to us.'

The Blitz itself began at 1633 on Saturday 7 September 1940. For fifty-seven consecutive nights German bombers unloaded the foul contents of their bowels on to the heart of England. Before the West End came under attack Soho crews were already in action in the East End and even as far away as the Isle of Dogs. But it was only five days before bombs struck the West End itself.

Of the two George Crosses awarded to British fire brigade personnel during the Second World War, one was to Harry Errington of Soho, the only London fire-fighter to be awarded the nation's highest bravery award during the Blitz.

Once the Blitz began, the men of Soho's Sub-Station Z in Rathbone Place found themselves working flat out day and night in all areas of the city. When they did get the chance to sleep they bedded down in the basement of Jackson's Garage. On the very first night of the bombing they went to Woolwich Arsenal, Peckham, Camberwell, East Surrey Docks, Great Eastern Street, and even chased a burning river barge down the Thames near St Katharine's Dock. On 8 September they stood by at Odhams Press, Long Acre (where there was another Soho sub-station), Clerkenwell and Shoreditch Fire Stations. On the 15th they fought a relatively small fire in Great Scotland Yard, and stood by at Farringdon Fire Station on the 16th.

The following night, exhausted to the point of nausea, they settled down as best they could in the basement of Jackson's Garage, which was dark, damp and warm on that late summer evening, 17 September 1940. As usual they had no beds to sleep on, just a blanket on the concrete floor and their tunics rolled up as make-shift pillows. Leslie Gardner was off that night: he usually slept between John Hollingshead and Harry, so they called over to their friend John Terry, who usually slept in the opposite corner of the basement, to come and join them at the Oxford Street end. Two firemen from another station had arrived to make up the numbers: they chose to sleep at the other end of the row. Against the opposite wall twenty local residents made themselves as comfortable as possible in the impromptu shelter.

Sleep was quick to wash over their spent bodies despite the distant rumbling of high explosive bombs landing to the east. They did not even wake as the bombs landed closer. And then closer still. Suddenly, like a bolt of lightning from hell, a high explosive crashed through the flat roof of the garage four floors above their heads and exploded on the ground floor, demolishing the entire building. Harry Errington came to to find himself standing in a fireball so intense it ignited the oxygen in people's lungs nearby: 'I woke up, standing on my feet, in the middle of the basement: I had been blown some twenty feet across the room. There was a great searing fire all around.' The appliances, taxi cabs, trailer pumps and numerous cars, their tanks filled with petrol, along with the entire garage petrol store, all crashed with a great roar into the basement, washing everything and everyone with a huge wave-like wall of flame. The entire building had collapsed like a deck of cards, killing all the civilians and seven of the firemen, including the two unfortunate stand-bys who had arrived only hours before. Harry stood still, momentarily stunned, then the pain of his burning flesh jolted him into action. He glanced in the direction of the staircase, but the way was blocked with sizzling rubble forty feet high:

Then I remembered that there was another exit in the opposite corner, closest to where we had been sleeping. So I ran. It was a natural survival instinct. I had somehow survived and I had to get out. As I reached the doorway, which was still almost intact, this new terrible sound pierced through my head, over the ghastly noise of the fire. It was John Hollingshead screaming. I picked up a blanket from the rubble, put it over my head and shoulders to shield me, and followed the sound back into the pyre. I found him lying face down, bare back in the furnace, with lumps of concrete burying him up to the waist.

Harry tore at the scalding masonry, cutting his hands, as his skin stuck to the hot rough surfaces and peeled away from his fingers and palms in long strips:

I got him up on his feet. It was like daylight, so brightly did the fire burn. As I carried Hollingshead towards the door, I suddenly saw John Terry lying down near the wall where we had been asleep. He had a large radiator across his feet and blood pouring from a wound on the side of his head. I went on with Hollingshead out into a small courtyard behind the shops in Oxford Street. I then went back to get Terry. Having moved the radiator, I got him to his feet, but he was so groggy he could hardly move. I put a firm hand under his armpit and virtually frog-marched him out. Finally I managed to get them both out into Oxford Street through a large window that had been blown out in the blast. We sat Terry on the side of the road where several first-aiders had already arrived and they took care of him until an ambulance arrived. He was taken from there to the Middlesex Hospital which was only about 250 yards away. But all around the area there were large signs saying 'Walking wounded to the Women's

Hospital, Soho Square' and this was impressed on my mind. I turned to John Hollingshead and said, 'John, come on, we're going to walk.'

I put my arm around him to take his weight, but he shouted in pain. I had forgotten that he had lost all the skin on his back. It was a terrible sight, but I knew he had to get to hospital, so off we set. We had taken a few steps when the air was filled with the horrifying whoosh of another incoming high explosive bomb. 'Get down!' I shouted, as we flattened ourselves against the road. I was face down in the gutter next to a pavement curb, pushing down into the ground, and I felt as if the curb was covering my whole body. The bomb hit John Lewis with a throaty thud, and once the dust had settled we went on our way. The staff in Soho Square were wonderful and treated us both for our injuries.

Five fire-fighters sit in casualty, waiting for treatment. The two on the left are LFB men, identifiable by the numbered badges on their tunics, the others are AFS. Eye injuries, cuts, burns, broken limbs and shrapnel wounds were all common.

The following morning Harry was moved to the Middlesex Hospital and then on to a hospital in Aylesbury. After a month of treatment he was discharged and headed straight back for his home, Soho:

My family now lived in Ridgmount Gardens and I made my way there across the West End, which had by now been widely bombed. My parents had moved to Birmingham and that evening just my cousin and I were

Harry Errington, GC.

there. We discussed whether we should go down into the basement when the sirens sounded, and decided that we should. Soon after the raid had started we heard a series of explosions getting closer and closer. Suddenly there was a great thunderous explosion, the whole building shook and we were showered with dust. The flats above had sustained a direct hit. We clambered out of the rubble and made for Goodge Street underground station. We went down on to the platform, found a small free area of platform and bedded down. We were so tired we awoke the next morning to find the rush hour already in full swing with busy trains unloading crowds of people all around us as we slept. That day I moved up to Birmingham and did not return to Soho until the end of the war. Months later I received a letter from the Home Secretary informing me that I was to be awarded the George Cross for my actions at Rathbone Place. It came as a complete surprise as I had discussed it with few people. I learned later that John Hollingshead and John Terry (later knighted for his services to the British Film Industry) had reported the events of that night to a senior officer.

The following night, the 18th, the two-man crew of a turn-table ladder were at work projecting a water jet on to a range of roof fires in Great Portland Street. The ladder was fully extended to a hundred feet and the fireman at the top was about to secure himself to the ladder by a safety belt when a bomb whistled down past him and exploded in the street directly below. There the second fireman was operating the ladder controls; he and an officer standing close by were severely wounded – the latter, District Officer Leonard Tobias, died later that night. But the fireman at the top of the ladder had what must be one of the most spectacular lucky escapes of all time.

When the bomb burst, the chassis of the turn-table ladder blew sideways into the front of the building, the rear wheel axle and ladder base section, weighing over four tons, blew right over the rooftop and came crashing to rest on another roof nearby. Various sections of the ladder were thrown also on to the roof, but the uppermost ladder extension caught on a projection, turned over through 360 degrees and hung down over the front of the building which was still on fire. The unfortunate fireman was thrown into the air, then fell to the pavement to be buried by falling debris. When the dust had settled a search was mounted in the area and the fireman was found still alive, but seriously injured. After months of hospital treatment, he survived the war and although disabled was able to run a public house in King's Cross, not far from the scene of his amazing good fortune.

Within a few weeks of the disastrous loss of life at Rathbone Place,

Sub-Station Z was relocated in Berners Street, next door to Sanderson's wallpaper warehouse. A number of new recruits were dispatched from the area training centre at Hugh Middleton School, Clerkenwell, to replace those lost. One of them was Giles 'nine lives' Harvey. He had only been in the fire service seven days when the training centre at Clerkenwell was demolished by a direct hit, which he survived against all odds without a scratch:

The crew were very shocked by the losses at Rathbone Place but we soon got our spirits back up, mainly with the help of old Hartman, a music hall actor known as the Drunken Wizard. He was such a ball of fire and fun that if Hitler had seen the laughs we used to get up to he would have given up! One evening I was duty man with the grand title of 'Standby-Garage', which meant I had to stay at the station co-ordinating fire appliances from other stations coming to stand by at Berners Street while the rest of the crew went out on shouts. There was a hell of a raid going on and I suddenly heard a couple more clearly than the rest: another one with my name written on it. One landed on the other side of the street and failed to explode. The other completely demolished Sanderson's building next door. The exterior of the station was damaged, of course, and we lost all the windows, but I was unhurt.

The next one that nearly got me was at a Royal Navy store in Deptford. After we'd been fighting the flames for hours, we were relieved and told to get a cup of tea. We had to walk for about fifteen minutes, but it was worth it. A wonderful welcome hot cup of tea (sometimes firemen would become so dehydrated that they did not pee for days on end) served by two lovely ladies smiling through the horrors of it all. We drank every last drop and set off back to the fire, but hadn't gone more than two hundred yards when a high explosive bomb made a direct hit on the canteen. It, the ladies in it and all the laughing customers around it were blown to smithereens, literally. We ran back, but couldn't even find any bodies. Terrible it was, terrible . . .

One night two young sisters, the Misses Teulon, reported to Soho to find out where they were to be based on their first night. They were detailed to Charles II Street Sub-Station in the basement of the Canada Life Building. No sooner had they arrived than a raid started. The building received a direct hit. Jessica Underwood, who was by now an Assistant Group Officer, set off immediately with Group Officer Lott down Shaftesbury Avenue towards Piccadilly Circus:

It wasn't much fun walking down the street in the middle of a raid, because there was so much metal and glass flying around. The shrapnel from the guns in the street was bad enough, but we always used to worry about what they were shooting at. I mean if they hit Jerry, he weren't going to be particular fussy where he come down, was he?

At Charles II Street they made their way into the basement where the air was thick with dust and rubble. In the midst of it all the two sisters were typing away by candlelight, carrying on as if nothing had happened. They refused to go back to Soho Main Station, arguing that just because it was their first night, it didn't mean they couldn't take the pace: they had had their baptism and they were going to see it through to the end.

Even in the bleakest moments of the burning of Soho there were glimmers of good humour and fun. Georgie Phillips, based at Odhams Press in Long Acre, recalled a typical event:

One night we had a serious fire in a gentlemen's club in Pall Mall caused by an incendiary which had landed on the roof tiles, burned through as they so often did, and set it well alight. It was going quite well on the upper floors, and one gentleman member, obviously a bit the worse for wear, tottered out into the street. He said he wanted to tell us what a bunch of heroes he thought we were and he kept asking if he could help in any way. Well, he wasn't in any state to help us, so one of the lads shouted at him, 'Give us a song and a dance', so there and then, in the middle of Pall Mall, he gave us a song and a dance, dressed in his long coat and bowler hat, fire burning all around.

On 24 September 1940, there were 740 separate working fires in London as a result of bombing, and many were in the West End, including a hundred pump fire in Southampton Buildings and large fires in Tottenham Court Road. In addition there were calls to more than one hundred serious fires in and around Soho, but the crews were so overstretched that sixty of them went unattended.

Jessica Underwood was on duty at the area control room adjacent to Euston Fire Station that night. News came in of a serious fire in and around Bloomsbury Street and Streatham Street. 'That's my home!' Jessica realized. Her mother, who was almost eighty, was at home in Streatham Street. Jessica set off into the dark night, her helmet on her head, and shrapnel falling to the ground from the heavy ack-ack shells exploding high above. She ran all the way down Gower Street and then cut right towards the flames. On the opposite side of the road from her home, the building was ablaze from top to bottom. The street was completely empty apart from a lone policeman who looked at the AFS badge on her helmet in complete disbelief. 'Blimey! Are you all they could send?' She was so breathless she could not say a word:

I made my way towards my mother's flat. Thank God it was still there. I dashed in, still very worried for her safety. She looked up surprised to see me. 'Oh hello, dear, we were just having a cup of tea, and wondering how you were getting on.' There she was with one of her old friends, having a cup of tea, and carrying on as if it was a summer's afternoon in 1920.

On 17 October, an intense incendiary and high explosive bombing attack took place in and around Oxford Street. Several large department stores were set alight and electricity, gas, sewage and water mains were all damaged; water pressure dropped dangerously low due to the excessive demands being made on the system from hydrants feeding greedy pumps all over the West End. A fifty-pump fire was in progress at Selfridges where the two top floors were alight from end to end. Superintendent George Benson, mobilized from Manchester Square Fire Station, ordered a water relay to be laid from the Serpentine in Hyde Park all the way to Marble Arch, and for hoses to be run from there to supply street water dams set up along Oxford Street. By then it looked as if a large part of Oxford Street towards Tottenham Court Road would be destroyed by the raging fire, but once reliable water supplies were available firemen were able to contain and surround the fires. All were under control by the next morning, although by this time both Selfridges and John Lewis' stores were gaunt, blackened ghosts of the fine buildings they had been the previous day.

As xenophobia spread through the very heart of Soho, leading to the internment of thousands of Italians and German Jews, Soho Fire Station swelled with foreigners. They counted Canadians, Americans, Poles, Norwegians and Danes amongst their ranks, including soldiers attached for fire-service training before going on into Europe. Meanwhile, the *News Chronicle* – a London-based daily paper – organized special weekends away with people in the country. The organizers visited all the stations and asked the men if they wanted to have their names put on the list. 'Benny' Goodman will never forget the weekend he took:

My name came up on the list and I fancied a weekend of not sleeping in Piccadilly Circus underground station with four thousand other Soho residents, so I set off to stay with Miss Sybil Cropper near Godalming. When I arrived the door was opened by a maid. Blimey! I'd never been to a house with a maid before! Miss Cropper made me very welcome and

Overleaf, top left: The blackened shell of John Lewis, Oxford Street.

Overleaf, bottom left: The noticeboard states the obvious as the crumbling walls that remain of John Lewis hang at a precarious angle. A trailer pump is set in to the hydrant on the corner and, to the right, a group of firemen continue to play their hose on to a still burning bank.

Overleaf, right: Four firemen work around a radial branch holder during the fire at the John Lewis building in Oxford Street. The one-and-a-half-inch nozzle was capable of delivering 600 gallons of water a minute with a throw of up to a hundred feet, but the water pressure would often drop and sometimes the jet would fail completely.

OXFORD
178
THE
NATIONA
DANGEROUS
WALLS

after lunch I decided to go the local picture house. I wore my fire uniform, of course, and to my surprise the box office boy let me in for free, but when the newsreel came on I knew why.

Benny sat near the front, the breathtaking black and white flickering, flaming images of Soho's burning reflecting off his silver cap badge in the gloom, his face illuminated by the ghostly white flashes.

'It was a very funny feeling,' he said coolly.

One October evening the telephone in the watchroom rang ominously. 'Soho Fire Station,' the duty girl answered it. A calm, characterless voice was short and to the point, 'Purple!'

This was the signal from control that the forward observers of the Royal Observer Corps had warned of a possible air-raid. The watchroom operator telephoned her colleagues and officers upstairs to give them the alarm. There then followed the usual tense period of anticipation, waiting for the phone to ring again. Would it be 'White' for the all clear, or 'Red' meaning the worst. Suddenly the phone sprang to life. 'Soho Fire Station,' the girl answered coolly. She was a telephonist in the City during the day and, like so many others, served in the Auxiliary Fire Service at night.

'Red!' the voice warned her. Simultaneously the air-raid sirens in the street began to wail their shrill warning all over the West End – known affectionately as 'Moaning Minnie', the siren had a spine-tingling quality that sent shudders of terror through all who heard it.

In the office on the first floor the girls, including Jessica Underwood and Company Officer Bentley, automatically stopped whatever they were doing, put on their tunics and made their way down to the basement shelter. They were well drilled, disciplined and organized though the basement was not always a pleasant place to take cover because it was filled with rats. After the main Newport Place dwellings were hit the following year, the rats thrived amongst the ruins, and used to swarm into the back of the station. One night a girl went down into the basement to use the lavatory. She went, in the pitch dark, shut the door on her cubicle, sat down and ... there it was rustling around her feet. She had been told that rats went for the throat of their victims, and she went into a complete high-wailing panic, screaming, 'My throat, my throat.' The whole station was awoken by the calamity but it took the firemen some minutes to calm her down enough to get her to open the door.

It was exactly 1900 on 7 October 1940. The firemen were just changing shift and several of them stood around in the watchroom along with the girls manning the telephones while the rest went down into the basement. Only two men remained upstairs on the second floor. Auxiliary

Fireman Fred Mitchell and Station Officer William Wilson were both doing important administrative work and chose to ignore the warning.

At 1903 a stick of bombs fell on Shaftesbury Avenue. Across the road from the station Sub-Officer Ernie Allday was walking down Frith Street and was about to cross Shaftesbury Avenue. He heard the bombs coming, but didn't have time to take cover. The blasts followed a second later. He blanked out for a minute and then came to lying in the remains of the ladies' underwear shop, S. Weiss. The underwear sold in the shop was known as 'naughty but nice', like Soho itself. Several Canadian soldiers emerged from the pub on the corner, picked Ernie up and carried him back inside for a quick check-up, double whisky and a pat on the back.

Several of the buildings on the north side of the avenue had been flattened. But when Ernie came back out into the street he was horrified to see that Soho Fire Station had also had a direct hit. The dust was still thick in the air, and rubble forty feet high spreadeagled its way across the road and into the side of the Palace Theatre. The Gaumont News Picture Theatre still stood opposite the station, but the front doors had been blown clean off. Ernie ran across the road and gained access to the remains of the fire station via the shattered front door. Most of the personnel in the basement were already emerging, covered in dust and some very shaken, but otherwise unhurt. Several firemen were trapped in the watchroom, the entrance to which had been completely cut off by fallen rubble.

Peggy Jacobs was driving a brigade staff car along the Charing Cross Road when the bombs fell, but the urgency of her mission meant she could not stop. In fact the street was filled with such thick smoke and dust she did not even realize that the station had been hit.

George Phillips 'came on' to Shaftesbury Avenue from his sub-station, and they were one of the first crews to arrive. When they did, to their great surprise, they found that the Soho firemen had not only released their colleagues trapped in the watchroom, but they had hand-wheeled the appliances out of the appliance bays, the ceiling of which had held. All the machines – enclosed limousine pump, pump escape, turn-table ladders, two staff cars and two motorcycles – were saved. The rescue crews searched desperately for the two missing men, but it was two days before Wilson's and Mitchell's bodies were recovered.

After several days of re-organization, Soho Main Station was relocated at the sub-station site in Stukeley Street, just off Kingsway. The station remained here until mid-1942 when a temporary fire station was constructed on the site in Shaftesbury Avenue using the original basement, watchroom and appliance bays, but with a prefabricated first floor office, mess, dormitory and rest room. Unfortunately the builders forgot to replace the sliding pole, so as an afterthought one was

constructed on the outside of the station. Once the station was back on the run, the external pole would cause much alarm in the pouring rain as 'speed of descent was near maximum velocity'.[1]

The destruction of property was deeply distressing for the firemen and women of Soho: the magnificent buildings – houses, churches, shops – were like old friends whose faces had been slashed in endless mindless attacks, but somehow everyone got used to it. It was the terrible loss of life that none could ever come to terms with. Like so many of their colleagues, the firemen and women of Soho carry forever the haunting memories of the human lives that the 'miserable bombing wrecked'. On 26 October 1940, St James's Residences in Brewer Street was flattened: five were killed and forty-five injured. Georgie Phillips' face was drawn and sad as he remembered the horrors:

By November 1940, parachute mines started falling (these were high explosive sea mines dropped on parachutes, which meant that, unlike the

Opposite: Shortly after the dust has settled, firemen dig frantically at the rubble of the fire station.

Above: Reinforcing appliances from adjacent stations arrive. Crews perform a hazardous search and rescue operation at first floor level. The Avenue Bar and Shaftesbury Theatre are clearly shown. The station doors have been blown off in the blast; the station obviously came close to complete destruction. The two appliance bays are unaffected, as is the watchroom, whose arched window is to the right of the gaping doors. Part of an advertising banner for the Palace Theatre opposite hangs precariously above the doors.

[1] This 'temporary' structure remained in use until 1982 when the present station was opened.

ordinary high explosive bombs which tended to penetrate into the ground, they exploded above ground level, causing much greater damage). One evening a mine was already on its way down when the siren started up. The Dominion Theatre, near the corner of Tottenham Court Road and Oxford Street, was turning out just at the same time. Opposite there was a very popular Lyons Corner House, and next to that a pub that was full of people. The mine came down smack in the middle of that lot! I have never seen so many dead bodies in my life: they were piled up on top of each other. In the semi-darkness I ran over to a windowless car. The occupants, a man and a woman, had had their heads cut clean off. I can see it now as clear as if it was yesterday.

Many churches were badly damaged or even completely destroyed, but one had a narrow escape: 'On 19 November 1940 a huge bomb crashed through the roof of St Patrick's Church in Soho Square, where it hit a column and buried itself in the nave but, miraculously, did not explode.'[1] Two magnificent churches by Wren were destroyed within a week of each other. The first, St Anne's, Soho, was hit on 7 October, the second, St James's, Piccadilly, on the 14th. The verger and his wife were both buried under tons of fallen masonry and died despite a twelve-hour struggle to release them. While this rescue attempt was in progress the

[1] Judith Summers *Soho* (Bloomsbury, 1989)

11 October 1940: Soho's burning! A lone crew get to work in Piccadilly, but their task is almost hopeless. They concentrate their jets on the buildings opposite, trying to stop the fire jumping across the street.

Carlton Club in Pall Mall was hit. Almost the entire Cabinet and Harold Macmillan were dining at the time in the room below the library where the bomb exploded. They all escaped without injury. After the war, the club moved to St James's Street. In 1990 it was the scene of great heroism by Soho's firemen when it was bombed again, this time by the IRA.

As the war progressed, the local Civil Defence Authorities decided to use bomb sites as temporary water dams. The supply of water to fires was one of the major problems of the Blitz. Water mains were often damaged beyond repair, so the artificial lakes were created with concrete and a central sump into which a suction hose was placed. They were fenced all around to keep people from falling into them at night. In the winter months, the cold weather made things worse, as the dams would freeze over and firemen were forced to crack the ice with axes and sledgehammers to enable the pumps to get to work.

Even at the very height of the Blitz Soho was still filled with prostitutes, also known to the men as 'Toms'. On a busy night you might count as many as forty or fifty between Cambridge Circus and Piccadilly Circus, lined up all along Shaftesbury Avenue. One evening the doorbell rang at the station. There stood a rather distraught-looking lady of the night. Hurriedly she explained, 'Oh, boys, please 'elp me. I've been using the dam across the road as me place of work. Just now, whilst I was doing me business with an American gentleman, I knocked me bleedin 'an-bag into the wa'er and it's got fifty quid in it. Can you boys get it back for me?' Now that was a king's ransom in 1940 – the auxiliary firemen were on the grand salary of three pounds per week. Some of the men went fishing with a long pole and a hook but had no luck. 'Another prostitute sneaked into a dam in Covent Garden, took off all her clothes, folded them neat and tidy on the side, and then just jumped in. Drowned herself. Sad really. We had to fish her out of course.'

'On the night of 8 March 1941, the basement Café de Paris in Coventry Street, advertised as "the safest place to dance in Town", was crowded with society people and off-duty officers, who were dancing to the sound of Ken "Snakehips" Johnson's band. "Snakehips" himself had spent the earlier part of the evening with friends at the Embassy Club in Mayfair. Determined not to let the Café de Paris's management down, he ran through the blackout, arriving in Coventry Street at exactly 9.45, just in time for his band to go on. The band struck up "Oh Johnny!", and more revellers crowded on to the dance-floor.'[1] On the streets above, the air-raid sirens were sounding, but none of the dancers even heard the warning. The bee-like drone of the bombers was followed rapidly by the

[1] Judith Summers *Soho* (Bloomsbury, 1989)

St Clement Dane, the Strand, 10 May 1941: another beautiful church is destroyed by the Blitz.

screaming shriek of five high explosive bombs falling in a line across the West End. The Gaumont Film Company in Wardour Street was hit, as was a large fur store in Oxford Street. 'Two high-explosive bombs had hit the Rialto Cinema above the club, and crashed down into it. One exploded at ground-floor level, the other burst on impact.

'The blast was devastating. Many people had their clothes torn off them. At some tables entire parties were killed. Strangely, in the midst of the carnage, a single champagne bottle survived intact. Reeling with shock, women and men in tattered evening dress stumbled out into the street, ghostly apparitions, their faces and bodies coated with dust, plaster and, in many cases, blood.'[1] There was a dreadful delay in the ambulances arriving: it took them nearly an hour to reach the scene. In the meantime first aid was administered in the Mapleton Hotel. Over fifty people were critically injured. Passers-by assisted however they could while others searched desperately amongst the injured for their missing loved ones and friends. Two nearby buildings were set up as temporary mortuaries: the Honeydew Restaurant and the premises of Skiggs and Russell.

In all thirty-four people were killed outright in the Café de Paris, including Snakehips Johnson, and dozens more were to die later as a result of their injuries. To make matters worse, exactly thirty seconds after the bomb exploded at the Café de Paris, another fell directly on the Madrid

[1] Judith Summers *Soho* (Bloomsbury, 1989)

restaurant just round the corner in Dean Street, killing another seventeen people and seriously injuring five more.

Another terrible attack came in the early hours of 17 April 1941. At 0310 a land mine devastated part of Jermyn Street killing seven people and trapping and injuring another twenty-three. The crew of Sub-Station X in Charles II Street were fully committed and called for assistance. Word then came that Admiralty Arch had also been hit but that casualties were light. As the rescuers continued their battle to save lives, the raid raged overhead. The combined noise of the anti-aircraft guns firing, the menacing growl of the bombers' engines high above, the whistling and shrieking of the falling bombs, the roar of the fires, the screams of the injured, the cries of the anguished all came together in a ghastly symphony of destruction.

At 0325, less than half a mile from Jermyn Street, the block of tenement flats in Newport Place immediately behind the main fire station was struck by a parachute mine. These flats had been built at the end of the last century principally as accommodation for the workers from Covent Garden fruit and vegetable market. A large part of the building collapsed, leaving forty-eight dead and eighty-three injured. There were many heart-tearing scenes of utter despair and desolation as a father would dig with his bare hands until he uncovered the remains of a small child in the rubble, twisted, broken, limp. One man was standing on his second floor balcony

A view of Tottenham Court Road taken on 24 September 1940, showing the broken water mains and other damaged supply pipes.

Above: Jermyn Street, looking towards St James's Street, the morning after a land mine exploded there in April 1941. Almost every building in the street is damaged. A lone fireman works his way through the rubble.

Opposite: The remains of the Newport Place flats, looking north towards the rear of the fire station in Shaftesbury Avenue. The chimney stack in the background is that of the Palace Theatre.

smoking a cigarette as the land mine exploded. He was blown through the air, still attached to the balcony, and landed without injury, but his entire family were killed: beloved wife, baby boy, little girl. 'You almost ran out of tears,' said one rescuer, remembering that night. The ugly side of humanity also showed itself when extra police had to be drafted in to stop the crowds of onlookers looting from the wreckage.

The final high explosive and land mine Blitz came on 11 May 1941 and Soho's ground took a particularly hard bashing. New Bond Street, Conduit Street and Bruton Street were all badly damaged. Carr's Hotel, Clarges Street, received a direct hit; a perilous rescue operation to free a woman pinned to her bed by debris was effected, whilst water was sprayed on a wall to keep the fire at bay. Five people were killed and three others badly injured. Mayfair Court, Stratton Street, was also destroyed – eight residents were injured and a number of BBC staff were rescued from the basement unhurt. Dean Street and Old Compton Street were badly hit. Twenty-four men, women and children were badly injured, of whom four died that morning; the buildings destroyed included the famous Patisserie Valerie which moved to 44 Old Compton Street – where it thrives to this day.

Once the first Blitz was over, the crews of Soho set to work carrying out vital maintenance work on their equipment, which had been in almost

continuous use day and night for three months. Hoses were repaired, water dams upgraded or built from new, engines maintained, hydrants repaired, drills improved, ladders replaced, recruits trained. In addition the crew were kept busy attending normal fire and emergency calls, as well as providing continuous cover in the event of further raids, but some of the fire-fighters grew restless for action and volunteered for service abroad; one of them was Ernie Allday:

A large number of us volunteered to go to Delhi and Calcutta, India, where they were expecting bombing from the Japanese. I served two years out there, came back to London and was called up within three months and just after I was married – the second front in Europe had started. After the first Blitz all firemen were de-reserved, so call-up papers for the services soon began to drop on to fire station office tables. I was ordered to the Army in the Far East. 'You're joking,' I told them, 'I've only just come back!' I refused to go and went before a tribunal, where I was given a lovely choice: the Far East or the coal mines of Yorkshire. I chose Yorkshire and spent the next eighteen months down the mines. Meanwhile my poor good lady wife was stuck in the control room at Clerkenwell Fire Station. As soon as I'd finished my time down the mines I transferred back to the London Fire Brigade, and this time I made it to Soho as Sub-Officer on the Blue Watch, where Hughie Abbit was Station Officer. The Red Watch were a fine bunch too, and their Guvnor was the legendary Ted Baldry. I was so pleased to be at Soho: the station had a wonderful character, and the ground and risk so varied I don't suppose there's another spot in the world like it. My Soho days were truly halcyon days.

A VI Rocket attack caused extensive damage in Old Compton Street on 24 February 1944. A gas main is well alight.

Bad weather was always a blessing as it gave the crews respite from the bombers' raids but the V1 fly-bombs, also known as doodle-bugs or buzz-bombs, and then the V2 rockets didn't care about the weather:

There was a very popular cafe in Colville Place off Tottenham Court Road. You could always tell a good cafe by the number of horses and carts parked outside it. The railways, milk board, vegetable and fruit traders all used horse-drawn vehicles. A V1 rocket landed in Colville Place and we set off before the bells had even gone down. We could see clouds of dust and smoke belching up from north of Oxford Street. As we got close, the air was filled with the sickening smell of freshly spilt blood. We turned the corner to find almost all the horses had been killed, while the rest were writhing about in agony and had to be shot. A young couple were running down the centre of the street, she clutching the limp rag doll like figure of their daughter. They refused to believe she was dead and insisted on running with her to the Middlesex Hospital.

Lit only by the flames, rescue crews, shielding their faces from the intense heat, move towards the broken ruins of a terrace in Old Compton Street to attempt rescues.

The flying bomb attack built up to a peak from June 1944 through July and August. The bloody buzz-bombs would sometimes arrive in waves. Each one made the sound of a dozen cars revving exhaustless engines on full throttle. Then the engine would cut out. 'As long as it carried on throbbing, you knew that you were OK but that some other poor sod was going to cop it.' When they landed after that dreadful silence, even the trees would be stripped of every leaf and branch.

The raids now came in the day as well as at night. One particularly unpleasant V1 attack occurred on the Aldwych on 30 June 1944. Dozens of people were killed and injured and many were rescued from the remains of their offices. By chance one of the London Fire Brigade photographers was in the Strand when the rocket landed. The images capture the first rescues by Soho and Clerkenwell firemen of the dazed and injured.

How was it that so many people were in the street and still working in their offices when this missile landed? The use of air-raid sirens had been abandoned when it became clear that the heart of London was continually at risk from missile attack – bombs arrived at completely random intervals and the emotional and physical strain on the general population of London as well as its civil defence and fire rescue workers was oppressive and damaging.

Pat Talbot served at Soho from September 1939 to 1956 and drove the brigade's first enclosed limousine pump. His enthusiasm, like that of so many of his friends, still bursts from him when he talks of his time in Shaftesbury Avenue:

I absolutely loved it at Soho. The place was filled with weird and wonderful characters whom I shall remember as long as I live: Sub-Officer Collins, Fritz Bernhard and Bill Povey, and of course the Guvnor, Ted Baldry. He was the most outstanding fireman I ever knew. He would never order a man into the job until he had made sure that the man or men would be able to get out quickly in an emergency. If at any time you had to 'stand by' at another station, when you reported to the officer in charge you would be greeted with 'Oh, you're one of Baldry's boys, that's good' and I am still proud at the memory. There was one chap called Dapper Coleman and he used to fix all the part-time jobs. He never took a cut of the money either, it was just his way. 'Strawberry' Griffen was another of the old school; he could never get enough strawberry jam slapped on to his bread. Charlie Stollery kept a locker full of pills and

With no prior warning, the landing of a VI rocket caused devastation in the Aldwych. The first jet gets to work while crews climb ladders to carry out numerous rescues. Passers-by rush to help.

Casualties, pouring with blood, are led away. A tree in the foreground has been stripped bare of its branches, and behind it the fascia of a building bears the scars of a thousand pieces of shrapnel.

Later that morning, firemen and a nurse carry out the grisly task of piling bits of bodies on the stretchers while a doctor carries a bag full of pieces.

BYV 321

Crews begin rescue operations at Warwick Court: the building has been cut in half by a VI, killing several residents; many others are trapped in the rubble. At great personal risk, the firemen work to release the victims. One casualty is seen being lowered to what remains of the first floor. A gentle hand protects an injured face.

ointment and medicines so if you felt queer or had a pain, he had the cure. One of the most notable characters was Shiner Wright.[1] He was a very religious man and every night before he turned in he would kneel by his trestle and say his prayers, asking God to give him the strength to forgive his oppressors. He got so much stick he took to saying his prayers in the toilet. I remember one night we locked him in, and at that moment we got a shout. The trouble was that it was actually a working job. Shiner was riding BA (the man assigned to wear breathing apparatus) that night, and we got into terrible trouble for leaving him stuck in the can!

Mick O'Callaghan taught me everything about the ladders. What he didn't know about them wasn't worth knowing. After I got my test I used to always drive the ladders with Mick as my Number One. On our very first shout it was raining, and as we were going down the Avenue I remember he kept on saying, 'Take it easy, take it easy.' But I thought I had everything under control, at least I did, until we got to the Circus. As I turned into Regent Street I felt her beginning to slip away from me, and the ladders did a complete circle and I ended up driving down Lower Regent Street against the flow of the traffic, which in those days just wasn't done. As Mick was one of the regular LFB men, he wore a round, peakless cap, similar to a sailor's cap. His favourite trick was during roll call, when the Station Officer wasn't looking, he would shake that hat off his head, roll it down one arm, roll it along the other arm, then back on his head.

By the end of the Second World War the day of reckoning heralded an horrific tally: Soho Fire Station's ground had been hit by a total in excess of 580 high explosive bombs; almost every major building within its precious square mile was damaged, including the British Museum and the National Gallery (hit six times). Piccadilly Circus had been hit twice, Leicester Square three times, Pall Mall six times, Horse Guards Parade and surrounding buildings twenty-one times, and Piccadilly itself twenty-two times. In addition a total of sixty-four high explosive bombs landed on or immediately adjacent to Shaftesbury Avenue.

Ten of the deadly parachute mines landed and exploded and one failed to go off, despite making a direct hit on the Palladium Theatre. Eleven V1 missiles exploded on the ground, including the disaster at the corner of the Aldwych and Kingsway. One exploded on the south side of Hungerford Railway Bridge and two more came down in the Thames. Only one V2 long range rocket hit and that was in Oxford Street.

Some of the wartime fire-fighters stayed on after the war was over, Georgie Phillips, Eric North, Sid Jacobs and Ernie Allday amongst them. Discipline was high in the service after the war: all of the men had seen

[1] All firemen surnamed Wright are known as Shiner: another old Naval tradition.

Cicely Courtneidge, the actress, talking to Roy Baldwin (centre) and Firemen Strickland and Hodges.

action in battle, either on the streets of London or abroad in the services. The Station Officer was second only to God himself. A fireman never even spoke to the Station Officer. If you had something to say to him you always did it through the Sub-Officer. Both men ate their meals in a separate room from the men.

Said Georgie Phillips:

After the war I rather fancied the job, so I applied to become a full-time member of the brigade. I spent all my time at Soho. 1939 to 1969. I loved it at Soho. With the Palace Theatre opposite we got to know lots of the stars: Thora Hird, Bud Flanagan, Vanessa Lee, Ivor Novello . . . they were all so friendly to us. When we got back from a call we used to pull across the avenue so as to back into the appliance bays. That meant you used to be right underneath the stars' dressing rooms, and sometimes they'd appear at the window and give us a wave, a wink and a smile.

One evening Norman Wisdom was appearing at the Palace. As his show progressed the audience roared with laughter and cheered and clapped their appreciation. Suddenly Mr Wisdom heard the bells of the appliances clanging outside in the street. He held up his hands to the rapturous audience, calling them to silence. Once he had complete quiet, the bells were quite audible to everyone in the vast theatre. 'Ladies and gentlemen,' he said, most seriously, 'your applause should be for them, not me!' The audience broke into even more thunderous cheers as a tribute to the brave men of the London Fire Brigade.

There was a spate of bombings in the early fifties when Soho was still the heart of gangster-land. There was intense rivalry between various families and businesses including bookmakers (both legal and illegal) and also between the various club owners. On one occasion a club was bombed in Frith Street. When the firemen arrived, they found people in need of rescue at almost every window right across the face of the building. Fast and efficient rescues were carried out on all floors and a number of people were in quite dire straits having taken in a lot of smoke. But none of them waited to be taken to hospital. Few even bothered to thank their rescuers. Instead they slunk off like foxes into the night. No names, no witnesses, no nothing. Not a word.

FIRE! 

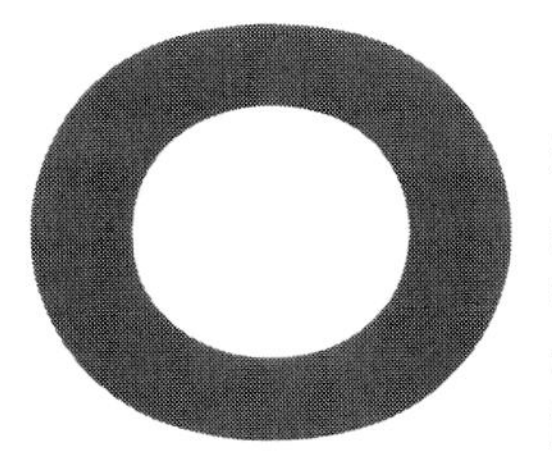

f all the rallying words whereby multitudes are gathered together and their energies impelled forcibly to one point, the cry of 'Fire!' is perhaps the most startling and the most irresistible. It levels all distinctions; it sets at naught sleep and meals, and occupations, and amusements. It turns night into day and even Sunday into a working day. It gives double strength to those who are blessed with any energy and paralyses those who have none.

It brings into prominent notice and converts into objects of sympathy, those who were little thought of or who were perhaps despised. It gives to the dwellers in a whole, huge neighbourhood the unity of one family.

Charles Dickens

FIRE! Like Soho, another four-letter word that evokes great passions: fear, desire, hate, adoration, obsession. Of all the forces in nature, fire is surely the only one mankind has partially tamed, yet it too has the potential to break free from the manacles we set around it, and devour all in its path. Having seen Moscow burn, the Emperor Napoleon wrote: 'It was the spectacle of a sea and billows of fire, a sky and clouds of flame: mountains of red, rolling flame, like immense waves of the sea, alternately bursting forth and elevating themselves to the skies of the flame above. Oh! It was the most grand, the most sublime, the most terrific sight the world ever beheld.'

The world is full of voyeurs, and nowhere more obviously than at a fire. People are drawn towards a fire, magnetized, then hypnotized by its splendour. A burning building, symbol of strength and nature's supremacy over man, is consumed in great greedy mouthfuls of breathtaking colour that serve to humiliate our pretensions. The marvel of it, the sheer terror of it, the wonder of it. No surprise then that crowds gather to point and stare: it is the best street theatre in the whole of London. We live in a world where sensitivities are dulled, our psyches bombarded with shocking

A hose-reel is used to extinguish a rubbish job in Horse and Dolphin Yard.

images of suffering: on the news, in feature films, in newspapers; but fire raises us above the level of insensitivity – it surpasses all other life events. The first-hand experience of a fire will lodge itself in a person's memory so richly that they will be able to recall every detail, every smell, every dance of waving flames, for the rest of their lives.

Given the correct circumstances almost anything you can imagine will burn: cars, piles of rotting rubbish, dead bodies, builders' skips, portacabins, derelict buildings, chip stands, ice cream vans, cranes, cookers, ships, lifts, conveyor belts, bundles of waste paper, scaffolding poles, computers, washing machines . . . In the last few years Soho's firemen have had to deal with all of the above burning and more besides.

The most common fire call is to rubbish, but even a pile of burning rubbish can hold within it the power to maim or kill. Station Officer Tony Wilmott commanded the Blue Watch at Soho for eight years and in all that time he only had one 'really nasty experience with fire':

We had been called to the wrong end of a small alleyway off Piccadilly. I walked down to have a look while the pump went around the block to reach the other end. As I got close I could see a small pile of rubbish alight, burning quite well. I took two steps closer towards it and an aerosol canister exploded, enveloping me in a complete ball of fire. I never realized how quickly you can close your eyes when you need to: every part of my face was singed except my eyes. My eyebrows were burned off, as was the hair sticking out from under my helmet, which just goes to prove the risks you are taking every time you respond to a fire call.

The West End seems to breed a particular kind of arsonist, nicknamed the 'rubbish loonies' by the men at Soho. For the most part the damage they do is not as great as the inconvenience they cause, but occasionally rubbish loonies have got completely out of hand. One night in 1988 a man dressed from head to toe in black gave Station Officer Bruce Hoad and his Red Watch a real run for their money. Up until 2330 it had been a relatively quiet evening shift: the ladders had been out to a couple of automatic fire alarm calls in hotels on other stations' grounds, while the pump had picked up a 'Shut in lift' shout, and a call to wash down the roadway under a car leaking petrol. Just after 2330, the bells went down again, this time to a rubbish fire in Tottenham Court Road. A pile of cardboard boxes left on the edge of the pavement was burning merrily, but was easily extinguished using a hose-reel. Half an hour later the pump was out again, this time to rubbish alight in Oxford Street near the junction with Tottenham Court Road. Twenty minutes later, another pile of rubbish burned on the pavement in Tottenham Court Road, then again, and then again.

'Twice is coincidence, five times is a rubbish loony!' Bruce cursed, glancing suspiciously at the small crowd of spectators. 'I wonder which one of you it is?' Half an hour later the contents of a builders' skip containing a good deal of burning wood put up more of a fight against the hose-reel. Then, within minutes of their finishing with the first skip, before they had even returned to the station, the RT (radio telephone) sprang to life, ordering them to another skip fire, this time off Rathbone Place. At the scene of the fire a police officer took Bruce to one side and told him cheerfully, 'You'll be glad to hear you won't be having any more trouble from this joker tonight. One of our officers saw him lighting this skip and we've arrested him. He's been taken to Savile Row police station.'

'Thank God for that,' Bruce replied, 'or else the little sod would have had us up the rest of the night.' Back at the station they plotted the various fires on the watchroom map: they formed a clear pattern across the fire ground. 'If he'd carried on, he'd have torched Great Marlborough Street next!' Bruce joked as the crews made their way up to bed.

A fire officer with a particular interest in the fire history of Soho, Bruce always had a special affection for Great Marlborough Street: it had after all been the location of the original Soho Fire Station, which had in its turn provided one of Bruce's most illustrious predecessors, Dan Ivall, with a spectacular twenty pump fire. Bruce sat daydreaming in the darkness of his study, recalling accounts of that great fire back in April 1970, which he had heard so much about . . .

A tremendous explosion ripped through the basement and ground floor of a dress manufacturer's, occupying the building which until 1921 had been Station 72, Great Marlborough Street Fire Station. Within minutes the entire building was ablaze, with flames showing at every window across all five floors. Soho's pump was out on another shout, and Dutchy Holland was riding in charge of the pump escape, so he was first on the scene. The pumps crew heard him make Pumps Four over the radio and Station Officer Ivall turned to them and said, 'I hope he has got a Four, and is not going to make a fool of me.'

A few minutes later the radio crackled to life again, this time making Pumps Six and then a minute later Ten. It was quite unheard of to make up in such rapid succession without having even sent an informative message describing the building involved and the extent of the fire. The Guvnor looked very serious: 'God, he really had better have a good job on his hands.' Mr Ivall arrived a couple of minutes later to find, much to his relief, that it was going very well.

While some crews tackled the blaze from the front, others traced their route through the narrow streets of Soho to the rear of the building, climbing walls, negotiating narrow parapets and finally scrambling on to the flat roof of adjoining premises from which the full extent of the fire could be seen. From every window of the rear extension to the building, some two hundred feet in depth, the yellow flames belched, and through the shattered lantern light of the roof on which they stood, the angry red glow could be seen beneath them. Again the long haul of heavy hose as, threading their way through various alleyways, through adjoining buildings, up external iron staircases, the breathless men laid lines to provide jets of water to prevent the fire from engulfing the closely packed surrounding buildings. Soho must be one of the only places in London where flames still jump across the street, setting fire to unsuspecting neighbours, as they did in 1666.

As the numerous crews moved in on the fire for the final kill, Divisional Officer Arthur Nicholls saw a hose-line leading up an iron stairway into the top floor at the rear of the building.

'As I climbed, Station Officer Dan Ivall crawled out of the building, hung his head over the stairway and retched violently, then he turned and went back in. I went in after him. "Pull your crew out, Dan, the building's in danger of collapse."' Four floors below them, Station Officer John Keevash, from A25 Westminster, was standing on the roof of an extension at the back of the building when it suddenly collapsed beneath him and he fell with a great crack to the basement level, breaking one of his knee caps. Far above him Pete Holmes, Knocker White and Tim Alderman – Station Officer Dan Ivall's regular pump crew – heard his cries for help and, looking down towards the fire, made out his body amongst the smouldering ruins. Quick as lightning they carried out the most unauthorized of descents by using their fully charged hose as a fireman's pole . . .

Bruce Hoad snapped out of his daydream as his room was flooded by all the lights coming on, switched automatically by the bells going down. He glanced at his watch: it was 0340. Once again the pump's crew jumped into their fire boots, pulled up leggings, put on tunics and helmets, and hauled themselves on to the appliance. Bruce glanced at the call slip the watchroom man had just handed to him. 'FIRE, Great Marlborough Street.' He read out the dispatch message to his driver and crew excitedly, 'and you'd better rig in sets too,' he shouted over his shoulder to the pump's crew in the back. A two-minute zigzag dash across the very heart of Soho brought all three of Soho's appliances into Great Marlborough Street. Bruce's eyes sparkled in disbelief; a building in the course of reconstruction was 'going like a bastard'. 'Make Pumps Six!' He snapped out the order to his driver as the crews flew into action.

While the fire in Great Marlborough Street was still going on, news came that two other serious fires were in progress on Soho's ground, both in buildings undergoing reconstruction. Once the fire was surrounded and virtually quelled, Bruce noticed the same policeman who had spoken to him a couple of hours earlier. 'What the hell happened?' he asked.

The policeman looked embarrassed. 'I'm afraid we had to give the gentleman bail. We only released him half an hour ago.'

A massive search was launched, but to no avail. Obviously content with his eleven fires, and no doubt discouraged by the onset of dawn, the pyromaniac disappeared. The next night he was back, but was caught and this time not bailed after the second fire of the evening.

Soho receives hundreds and hundreds of malicious false alarm calls each year. The people who perpetrate these crimes are ignorant, despicable, sick deviants. Sometimes they just pick up a public telephone, shout 'Fire' and then slam down the receiver. Under such circumstances, the British Telecom operator initiates a trace of the call, now a very fast procedure. Once the location of the telephone from which the call was made is traced, the operator then telephones the fire brigade control, which in turn dispatches fire appliances to the address of the telephone. The London Fire Brigade simply cannot afford to ignore these calls, even if they are made by a young giggling child. One in a thousand of these 'abandoned' calls will be the real thing. A confused, perhaps foreign, child in a flat alone, caught in a fire, may think all you have to do is pick up the phone, dial '999', shout 'Fire', and help will arrive . . .

A recent trend has been for false alarm calls to extend to the bizarre: a call in the middle of the night to a 'Road traffic accident, persons trapped, the Strand'; a desperate sounding call to 'Two underground trains crashed in tunnel, many casualties, Oxford Circus underground station' at 1100 on a Tuesday morning in spring. The underground stations in central London are now fitted with push button 'alarm points', clearly marked: 'If you discover a fire press this button. No alarm will sound on the platform but the fire brigade will be called.' For some morons this seems to act like a red flag to a bull. Charing Cross underground station is a particular favourite for this little prank, often as many as three or four times in one night. Every time the bells go down in the fire station, the men risk their lives driving at high speed to respond to the call, in case it is the real thing. The stress of a turnout is high, the cost to the taxpayer is enormous, and the potential cost in terms of the lives of people who might be in genuine need of rescue while the appliances are tied up attending another false alarm is unquantifiable.

Whenever the driver of a fire engine 'jumps' a red traffic light he is breaking the law. He takes this risk upon his own shoulders and will bear the dire consequences if he makes a mistake. He knows that every second counts if the call is the real thing. A small fire can become a very big fire in two or three minutes. His job is to get his crew and their equipment safely to the scene of the incident in as short a time as possible. Most fire stations turn out about three or four times in one fifteen-hour night shift. At Soho they go out on average fourteen times: sometimes it is as many as twenty-five or thirty times in one night! The crews love the fact that the station is so busy – that's why they are at Soho – but they despise malicious false alarms, almost as much as they hate automatic fire alarms (except when the alarm is raised in good intent. Recently an over-enthusiastic thespian rehearsing his lines at the stage door, yelling 'Fire!' at the top of his voice, was surprised when all three of Soho's engines arrived at breakneck speed).

Automatic fire alarms – AFAs – account for a majority of the fire calls Soho attends. In the last few years a radical change of emphasis has taken place in fire-fighting strategy in London, moving from a response emergency service on the offensive, to a prevention-based defensive service. Fire prevention, fire certification, fire safety inspections are all now principal tasks of the fire service. Every hotel, multiple occupancy building, theatre, cinema, public building, office block and department store has an automatic fire alarm system of some description. The most common alarms are activated manually with 'break glass' buttons or automatically by smoke detectors. These high tech pieces of equipment are also variably activated by steam, cigarette smoke, mites, flies, burnt toast, vibrations, builders' dust, power cuts, lightning, electrical short circuits, dripping water and even, on one occasion in a very grand hotel, a nest of gnats! Many buildings also have wet and dry sprinkler systems: a network of pipes at ceiling level interspersed with glass nozzle heads which burst open to release water under predetermined levels of heat. These rarely malfunction, but it is not unknown.

The biggest problem with AFAs occurs when they go off at night in commercial buildings. Firemen spend hours waiting around for the police to locate and summon the necessary key-holder of a property where an alarm is sounding. If no fire or smoke is showing, the fire officer in charge is not allowed to break into a building simply because an AFA has gone off; he has to wait, sometimes for hours, for the key-holder to arrive and unlock. A system of fining companies who are regular 'offenders' has been discussed, but the maximum fine available is pathetically small, and it might also have the effect of discouraging people from calling the fire brigade immediately an alarm goes off.

Despite the disheartening number of 'good intent' false alarms caused by AFAs, they have, on occasion, saved the day. In 1985 the Blue Watch were called to an 'AFA actuating, the Savoy Hotel'. Station Officer Shiner

Wright was in charge. On arriving in the Strand he sent Bruce Hoad into the hotel to investigate:

We pulled up at the front entrance where we were met by one of the doormen, who said, 'Would you mind parking in a side street – you're blocking the taxi rank and your flashing lights will disturb our guests.' So I told him his fortune and continued into the lobby. It was Christmas night. Inside I was met by another member of staff who informed me that the alarm had gone off three times. The first two times they had cancelled it: the third time two members of staff had gone and had a look, couldn't see anything and so had turned it off again. People in that section of the hotel phoned the switchboard to ask what was happening only to be told, in true *Titanic* style, that there was no emergency and no need for alarm.

A Volkswagen camper van going well in St Albans Street.

At that moment their portable radios crackled into life: 'Guvnor, we've got a job going on the first floor. We're going to need sets and a jet.' Two of the crew had discovered a smoke-filled corridor.

Back in the entrance:

Once we realized that there was a fire in progress, we ordered the main fire alarm to be re-activated and when the duty manager said no, Leading Fireman Roger Kendall actually grabbed him by the neck and pulled him across the front desk, just to make the point that when a fire officer asks for something he means it.

We began to lay out hose and they had the cheek to ask us not to take it up the main staircase, but please to find another route. We asked to see the General Manager but they refused to get him. I screamed at him, 'Even at five hundred pounds a night they're still going to burn, you silly bastard. Get everyone out of this hotel. Now!'

Meanwhile the fire was burning fiercely in a store room on the first floor. It was going well and had smoke-logged the whole corridor, which had several bedrooms nearby. When asked, the management didn't have a clue which of the staff bedrooms were occupied and which were empty.

Leading Fireman Roger Kendall gets to work with a hose-reel: 'Dear God, let's hope there's no one in it this time!'

As a last resort to try and get the full co-operation of the hotel staff, the officer in charge had to threaten to summon the duty Fire Prevention Divisional Officer and actually have the place closed down. Most people do not know that the fire brigade has enormous power with regard to the enforcement of public safety measures. It would seem that most hotel staff are ignorant of this fact. Worse still, they often do not seem to care: 'Once we began fire-fighting and searching the rooms around the fire, we found a lot of the doors were locked. The management were not even able to provide keys to the bedrooms and we had to break down the doors!'

A spectacular fire at Villiers House, the Strand, 18 January 1981. The photo was taken early on in the job as the fire rolled in exploding balls from floor to floor – a real-life towering inferno.

Assistant Divisional Officer 'Speedy' Close was now in charge of the incident and said to the front of house manager, 'I want to speak to the man who is actually in charge of this hotel. If I don't, I will have this place closed down immediately and your guests will be looking for other accommodation.'

The Fire Prevention Officers were called to the scene and the staff of the hotel were read the riot act and warned of the consequences of their failing to enforce all the appropriate fire precautions and actions in the event of the automatic fire alarm going off.

In 1990 the Savoy's automatic fire alarm actuated again: this time they called the fire brigade immediately. Just as well: the Savoy Theatre was ablaze from end to end and in danger of spreading to the hotel next door. The Red Watch subsequently held the fire at a ten pumper, carrying out particularly excellent fire-fighting for which they received a letter of congratulations from their Area Commander.

In the summer of 1991 the Savoy Hotel had another serious fire: unfortunately the actions of the staff were most unsatisfactory – the first fire crews to arrive were taken to the opposite end of the hotel from the location of the fire. In addition the internal dry-rising water main system failed to function correctly and the fire had to be fought using twenty water fire extinguishers.

Of the more serious fires the most unpleasant and most common that Soho has are restaurant fires. They nearly always involve a kitchen in the basement and ducting from the kitchen to the top floor. Imagine what it is like to be a fireman fighting such a fire: you go into the ground floor restaurant which is so thick with smoke that you can't see your hand in front of your face. Every few yards you collide with tables and chairs, sending glasses and cutlery crashing down on top of you. And all the time you're dragging the hose with you: a feat similar to dragging a seventy-five-foot python that has been stuffed with concrete through an underground assault course maze. You can feel the fire and you can hear the fire. The floor is even beginning to burn from below. But the one thing you cannot do is find the entrance to the basement. It might be tucked away in the far corner of the room, or it might be behind the bar. The whole restaurant might only be thirty feet across, but when you're crawling completely blind in breathing apparatus, surrounded by swirling hot gases, distracted by the crackling of flames, the hissing of the charged dripping hose and confused by the loud banging of your heartbeat in your ears, it's easy to become completely disoriented. For the first crews into the fire it's like going scuba-diving in a deep lagoon of boiling water. Soho has literally thousands of restaurants and nearly all of them are located in buildings that were originally constructed as residential properties. The kitchens are usually filled with highly combustible materials including large stores of oil and sometimes Calor gas cylinders as well.

The Strand and its environs have provided Soho firemen with some of their greatest fires: Grand Buildings on the corner with Northumberland Avenue has seen one thirty, two twenties and an eight pump fire in its time. Station Officer Tony Wilmott, commander of the Blue Watch from 1976 to 1984 (although he had first served at Soho as a fireman in 1960) 'had two goes at burning down the Strand'. The first time was in January 1981. The building belonged to HM Customs and Excise and had slipped through the fire prevention inspection net on being sold to a business enterprise as offices. As so often in Soho, there was nothing showing at the front of the building, but Tony had only to glance down St George's Court, a narrow side street running from the Strand towards the river, to see great sheets of flame roaring out of several floors of the office block.

The rapid spread of fire was quite horrific: it didn't spread internally at all. It spread outside, jumping from window to window. Afterwards, Assistant Chief Officer Jim Stevens (also an ex-Soho Station Officer) said, 'Had that building been two hundred storeys high we would have lost the lot, because it was going faster than we could contain it. Thank God there were no people in the building at the time. It would have been a major disaster.' It took the crews of forty pumps to win the battle, by which time almost the entire building was burned.

Eighteen months later another enormous fire spread with dramatic speed through the Civil Service department store, Chandos Place (near the site of the original Chandos Street Fire Station. Chandos Street was renamed Chandos Place after the war to avoid confusion with the Chandos Street off Cavendish Square). The building was in the course of redevelopment and contained dozens of cylinders used for welding metal. At the height of the fire the cylinders began exploding throughout the building, and some even launched out of windows into the street like flaming rockets, exploding in the air or in nearby buildings, showering the firemen on the ground with shrapnel. It is probably the closest to an action replay of a Blitz air-raid Soho's men have experienced since the war.

Covent Garden has also seen many serious fires. In December 1949 a fire broke out in the cellars which stretched the full length of the flower market. In all 62,500 square feet were burned: crates, sacks, Christmas trees and other foliage. Thirty firemen were taken to hospital and Station Officer Fisher (rescued gallantly by Station Officer Taffy Watkins) from Whitefriars Station (just off Fleet Street, and at that time the 'take' station to the south east of Soho's ground) died having been overcome by smoke. He was married and had three children. He was a also a very popular, experienced Guvnor and a 'gallant gentleman'. Among the others injured

The charred remains of the Civil Service department store in the Strand, 27 July 1982. Soho's pump escape, on the right, drinks greedily from the hydrant water main, while simultaneously pumping water to various jets being used by crews on the outside of the building. The men have been withdrawn from internal fire-fighting because of the danger of collapse and exploding cylinders.

were Leading Fireman Wilson and Fireman John Hood from Soho. Mr Delve, the Chief Officer, said later, 'There was no vertical ventilation, so the smoke could only come out of the passage at the end of the building and through the chutes which we opened up, but these outlets were the men's only inlets.' It was two days and nights before the flames were eventually subdued.

In May 1954 the Poparts Warehouse was burned to the ground. During the fire-fighting operations three men were trapped in a collapse. Two died very quickly but the third fought for his life for several days before he too capitulated.

A fireman's first priority is the saving of life and to this end a good deal of their training is directed. The use of ladders and lines (ropes) is a central tool in the fireman's pot-pourri of knowledge. In the last forty years hundreds of people have been rescued from the fiery jaws of certain death by the firemen of Soho and many of these rescues have never been recorded officially. The valiant actions of others were not witnessed due to the rescues taking place at the rear of buildings out of sight of their colleagues, or else the rescues were played down by the perpetrators, never wanting to blow their own trumpets. 'There's no need to make a big fuss, it's what we're paid for, isn't it?' Firemen are by the very nature of their job reluctant heroes. Occasionally a deed of great courage involving actions above and beyond the call of duty is actually witnessed by another fire officer and passed on to the Honours and Awards Board – the team of fire officers who suggest fire-fighters for commendations and certificates in recognition of their actions – although it is often only once a full investigation into a fire has been undertaken that the full facts of heroism emerge.

One such incident occurred in 1949. Peter Jackson and Ernie Allday

21 December 1949: firemen struggle in a nightmarish basement below the Covent Garden flower, fruit and vegetable market, the steaming water in places coming up to their chests.

Crews of the London Salvage Corps, in distinctive helmets, stand in the foreground of the Poparts Warehouse, Langley Street, Covent Garden, just prior to the collapse in which Station Officer Fred Hawkins (of Clerkenwell Fire Station), Fireman Charles Gadd and Fireman Arthur Batt Rawden were all killed, 11 May 1954.

were part of the crew riding that night. For three nights running they had had a number of malicious false alarm calls to 'Smoke issuing' in the region of Regent Street in the middle of the night. Despite this, as a safeguard, it was standard practice at Soho to put the turn-table ladders up in the air to their full extent of one hundred feet, with a man at the top as a look-out to see if they could see anything from above. At 0534 on 2 February the bells went down again calling them to 'Smoke issuing, 10 Old Burlington Street'. 'Probably just another bloody false alarm,' they assumed, but remembering the golden rule, never judge anything by its cover when on the fire ground in Soho, up went the ladders as usual, just to check.

The TL man made the long climb to the top while the rest of his crew waited below. After a few seconds at the top he came leaping back down and ran over to the Guvnor, panting excitedly, 'There's smoke punching up from behind this building, Guvnor!' His eyes sparkled with excitement. He knew from the density and dark colour of the smoke that they had a good job going. (A 'good job' is a fireman's expression for a fire which presents them with a challenge: a working fire which taxes all their resources and skills to the very limit. Fireman are not callous pyromaniacs who long for disaster to strike the public whose lives and property they guard, but if a fire breaks out, and it's a good job, then the excitement and exhilaration they experience is akin to that of military commandos who have trained and trained for months to attack a particular enemy target.)

They broke into the building, and Ernie Allday led the way, running along the corridor and up the stairs:

We could hear it crackling and spitting above us. The fire was in the kitchen on the second floor which faced on to the stairwell. It was going well and the thick oily black smoke was really belching out, and of course we had no BA in those days, so you just got on with it as best you could and tried not to swallow – or yaffle as we used to call it – too much.

Suddenly Ernie Allday heard faint calls for help from above and managed to cut off the flames from the burning room by closing the blazing door of the room on fire. Then with Fireman Jackson he forced his way up to the third floor where, after a search by crawling, he found a man lying in a doorway in a state of near collapse. Meanwhile the pump's crew had followed up behind them with a branch which gave the kitchen a good 'wash-out', checking the fire and improving conditions on the staircase.

As soon as the fire was pushed back into the room, Jackson and Allday carried the man down to ground level using great presence of mind and determination. Some weeks later the gentleman had made a good recovery and asked several of the firemen to go and have a drink at his club. Whilst there he told them that he had been trapped for hours in the rubble of his previous Soho house during the Blitz: 'As I was lying there last week slowly slipping towards death I remembered that fateful day, wondering whether I was ever going to be rescued . . . and then suddenly there you were, out of the smoke . . . bloody marvellous! Cheers, boys!' He raised his glass high in the air, smiling from ear to ear. 'I'll never forget you.'

It had been a busy tour for the White Watch under Station Officer John Beauvenisor. On the second day duty they had had three four-pump fires in one morning, including one in a fine seafood restaurant in Leicester Street. All the kitchen staff escaped the flames before the arrival of the brigade, but John ordered his crew to carry out search and rescue operations immediately. Tim Alderman entered the burning kitchen and came upon a large lobster which was obviously in great danger: 'I picked it up and put it down the front of my leggings: the poor thing would have been cooked alive if I hadn't rescued it. But it bit me with its pincers as a thank you, so we had it for lunch!'

Early that afternoon Wembley Control received dozens of calls to a fire in the Odhams Press Building, Long Acre – the same ill-fated building that had been bombed during the First World War, killing thirty-one people. Tony Cooper was driving the pump escape:

I pulled up at the end of Shelton Street. There was a small fire showing at the first floor, so Pete Holmes pulled the hose-reel off and began to spray the flames. At that moment I glanced down the street towards Long Acre and fifty yards away a jet of flame jumped thirty feet out into the street. I shouted at Pete, 'You're going to need more than that, mate,' and pointed down the street. His face was a picture.

Station Officer Beauvenisor immediately made Pumps Six. The spread of fire was dramatic and the heat of the flames so intense that all the buildings in the adjacent streets were in peril. As the fire progressed, curtains of water had to be sprayed on to the surrounding buildings as the wooden window-frames across the frontages burst into flames due to the radiated heat. In the street, the smoke was appropriately ink-black to the extent that the crews operating the ground monitors were forced to wear proto breathing apparatus.

During the summer of 1962 Tony Wilmott – then a fireman – drove the pump escape to a fire in Wardour Mews. On arrival the crew found there was a lot of smoke showing and Sub-Officer Terry Spindlow, in charge that day, saw a woman leaning out of a window on the first floor. Because

An injured fireman, his face blackened and burned, is carried to a waiting ambulance having been rescued from the Poparts Warehouse.

of the neon signs hanging over into the street, the crew couldn't get the fifty-foot wheeled escape ladder down the mews. An extension ladder was pitched instead but as it came up towards the lady, she jumped and grabbed on to its underside. Terry ran up and held on to her with all his might as the ladder slammed down against the burning building. Pat O'Brien grabbed a first floor ladder underneath and reaching up, supported her by her rear. It was a chaotic, haphazard rescue, but eventually they brought her down to safety. The fire-fighting itself went well because the driver, Danny Coleman, was a first class pump operator. He was so fast they had a jet almost as soon as they pulled up. Water on to the fire knocks the flames back inside the window.

While automatic fire alarms have existed in large commercial and office buildings for years, it is only in the last few years that smoke alarms have been available for general use in domestic homes. Anyone reading this text who does not have smoke alarms fitted in his or her home is a fool. Soho's fire-fighters encounter endless little tragedies every day in their work: an old age pensioner whose modest but uninsured flat contents have been reduced to a smoking pile of ash, or a restaurateur whose life's work lies as a smoking ruin . . . The destruction of property in fire is dreadful for the sufferers, but all of these losses are as nothing compared to the endless, pointless deaths, almost always by smoke rather than flame; smoke that steals away lives so quickly, so easily, so casually . . .

One night when Ernie Allday was Sub-Officer in charge they got a call to Berwick Street, to 'Smoke issuing'. Up the stairs they flew until at the top floor, which was heavily smoke-logged, they forced their way into a small crowded bedroom where an old Polish gentleman was asleep, his bed burning around him along with a wicker chair beside it. They shook him awake vigorously and carried him out, half unconscious; miraculously he was still alive, although the smoke was as thick as fog in his room. They carried him down the stairs and out into the street where he made a remarkable recovery. The cause of the fire was investigated and it turned out the old man had left a candle on his bedside chair; it had burned down to its base, caught the seat alight and then the bedding.

In the weeks after the fire he sat outside the shop watching the world go by, chatting and joking with the street market stallholders of Berwick Street as he had done for years. Whenever any fireman passed by, he would acknowledge them rapturously, telling everyone at the top of his proud, thick-accented voice, 'Those boys saved my life.'

Several months later the Blue Watch received another call to 'Smoke issuing' at the same address. Station Officer Hughie Abbit was in charge. As they arrived, Ernie called to him, 'Oh, I know this one, Guvnor!'

Abbit looked surprised. 'What do you mean?' he asked.

'I've been here before. You're going to find an old man up there.'

The Polish gentleman had done exactly the same thing all over again: the bed was alight and the chair beside it was burning too, but this time one extra thing was afire. They dragged him out into the street but they were too late: he had already gone.

In the late 1950s a serious fire broke out on the third and fourth floors of a very grotty premises just off Shaftesbury Avenue. Station Officer Dan Ivall, a legendary character at Soho and known as 'Lord Shaftesbury', was in charge. On arrival he was told that there were a man and woman trapped on the upper floors and a man at the back of the building hanging out of a window. Ivall ordered a crew in the front door, including Fireman Roy Baldwin who went up the front stairs and located two bodies in the

Fire has burst through the roof of offices in St George Street, off Hanover Square. A fireman prepares to enter a second floor window from the fifty-foot wheeled escape ladder. The blaze in the room to his left is attacked with a jet of water from the top of the turn-table ladder. Officers assess the overall spread of the fire from a hydraulic platform cage raised up to the height of the roof.

bathroom of the burning flat: 'It was so hot I thought I was going to die. I picked up the woman, putting my hands under her arms, but to my horror she slipped straight through: her skin peeled off in my hands like gloves.'

In the meantime, Ivall had gone around the back of the building where there was a well. He took with him a hook ladder. It was small, manoeuvrable and exactly what he needed. High above him a man was hanging by his fingertips from a windowsill on the fourth floor, with flames licking out around him. Without hesitation Ivall unclasped the large hook, made of steel with a serrated edge, clasped the wooden ladder firmly in both hands, held it vertical, slammed it against the first floor windowsill, hooking the hook over the edge, and began his climb. In less than a minute he had scaled the back of the building, reached the man and brought him to safety. It was a textbook rescue and exemplified the need for hook ladders on Soho's ground: no other ladder would have enabled the man to be rescued.

Fires are always a serious business but even at the height of a life threatening event, comic situations arise: in 1971 a fire broke out in the basement of the 'Israeli Shop in London'. It was particularly fierce and difficult to fight as the staircase from the basement emerged in the middle of the shop on the ground floor. Stacy Waddy, a journalist, was living on the fourth floor with her husband at the time:

You really don't think it will happen to you. That the West End's Biggest Blaze will be you. Your house. Your clothes. Well, it wasn't exactly a blaze, more of a malevolent smoulder which consumed half the house and took five hours to put out, and what it didn't burn, like the Stones song, it painted black. We lived on the top floor of an eighteenth-century rabbit warren in New Oxford Street. Four a.m., and my husband says I woke him sleep-talking about Smokey Robinson and the Miracles. He woke up enough to realize it was real smoke and started downstairs to see if I'd left the iron on, or if the insomniacs downstairs had set fire to some baked beans. He could only get down half a flight. I picked up the telephone and had dialled the first '9' before all the electricity failed. We put on what clothes we could grab in the dark and suddenly understood what being 'overcome by smoke' means. Somehow when other people are described as 'overcome' it sounds as though they were incompetent, or drunk, or doing something heroic, though possibly unnecessary.

We were already very nearly overcome. Thank God we had a balcony, for between waking up and climbing out on to it we had about three minutes. Already the whole flat was impenetrably filled with smoke. From the moment the fire engines came – twenty in all – it seemed absolutely normal. Pregnant as I was, and on the fourth floor, there seemed nothing a bit worrying or unusual about shivering on a balcony, first waving till they saw us and then waiting for the ladders. The first red wooden one was too short, so we waited while an immense chrome ladder, a masterpiece of articulation, swivelled and stretched and clicked its joints till it met us. The fireman at the end [Leading Fireman Ray Worboys], peering at our long hair in the smoke, reported over the intercom to Tony Cooper, who was operating the ladders, 'I've found two females on the fourth floor.'

I was more incensed by that than scared by everything else going on. 'That's my husband!' I yelled in fury and this amazing, calm, strong man said, 'Correction: One male, one female' into his microphone and helped me out on to the ladder. He held me in a grip that would stop a drowning wrestler from flailing – or feeling afraid – all the way down those miles and miles of sharp chrome bars. It always took my breath away, coming up the ordinary stairs. My husband seemed to get farther and farther away, a pale face in the smoke.

It was amazing how many people seemed to be about in New Oxford Street at 4 a.m. Lights popped on in windows we'd always assumed were offices. A magnificent neighbour made tea throughout the night and all around there were battalions of people in pyjamas and dressing gowns, telling us about condemnation orders and preservation plans, and reporters asking about who was rescued from where and when. The reporters had a field day: a parrot was given oxygen in an ambulance; the two flats downstairs were described as a 'commune'; a film director – from the third floor – was said to have left with two 'models' and we were described as 'Miss Stacy Waddy, a journalist, and a man'.

I was feeling a bit sick by now, so I asked the police for help. As the baby seemed to be churning about a bit we went and slept for a couple of hours in a detention cell in the police station. I felt, persistently and longingly, that I just wanted to go home. We knew the flames hadn't reached the fourth floor and somehow I imagined that we could leave the blackened wreckage downstairs in the foam (which had been used for the very first time) and just go to bed. I never thought that there would be no water, no electricity, no sewage. We climbed up the pitch-dark stairs, crunching broken glass, falling over smashed doors and debris, still thinking we were about to reach our sunny white place. We came into it and it was black. Every single thing, black. It looked like the idea Antonioni used for a fruit stall in *Red Desert* and John Boorman used for Notting Hill in *Leo the Last*: painting everything black to convey despair. And everything was – the ceiling, the walls, the floor, the banisters, the shelves, the curtains, the bedding, the rugs. It was almost a work of art now, where the onlooker recreates from the white circles and silhouettes an eerie third dimension of what once was there. It could have been worse, it's true. We told a friend the story and he kept saying, 'It could be worse' until we asked how and he said, 'It could have happened to me!'

On John Peen's first night at Soho, 16 July 1974, he picked up an eight pump fire in the crypt of St Martin-in-the-Fields, followed almost immediately by a ten pump fire in St Martin's Lane. Several months later the White Watch carried out multiple bird rescues at a particularly complicated fire at a film studio in Floral Street, Covent Garden. Two parrots were carried out, followed by a very angry toucan, whose huge beak snapped at Eddie Martin like a crocodile.

Most firemen will only be directly involved in a rescue once or twice in their entire careers. Usually only one or maybe two people will need to be rescued; at worst, perhaps an entire family. Very few firemen will ever be confronted with the daunting prospect of 'multiple rescues'. One such fire occured at the Worsley Hotel in Maida Vale in December 1974. Thirty people were rescued from all floors of the blazing building down every ladder available to the men on the scene. A full account of this fire, entitled *Red Watch*, was published in 1976 by Gordon Honeycomb. Several of Soho's Red Watch who were there that night received Letters of Congratulations for their actions at the fire in which one fireman was tragically killed, along with seven other people who had been in the hotel.

Several years later the White Watch at Soho picked up a multiple rescue job on their own ground in James Street. Covent Garden was still filled with old warehouse properties, previously used to store fruit and vegetables for the market, but now empty and awaiting redevelopment. Access via Long Acre was very difficult because of parked cars on double yellow lines, and James Street itself is narrow. It was ten o'clock at night when the bells went down. Under normal circumstances, once a serious fire shows itself in the street, the fire brigade control room receives many telephone calls to the same incident. When this happens the control officer manning the radio telephone calls up the appliances already on their way to the incident and informs the officer in charge that control is 'receiving multiple calls'. As soon as this message is broadcast, all who hear it know that the call they are proceeding to is going to be 'the real thing'. It gives the officer in charge extra warning of the severity of the fire which he and his crew are approaching.

Station Officer John Peen glanced at his driver as they turned into Long Acre. The radio was silent. There was no sight or even smell of smoke. Then they turned the corner into James Street. JP drew in his breath sharply in surprise at the spectacle his eyes struggled to take in. Across the entire face of the building, men and women were screaming for help from every single window on all three upper floors:

There were so many people to rescue it was almost impossible to know where to start. I made Pumps Eight, Persons Reported, and shouted encouragement to the crews as they sprang into action. Every single ladder was stripped off the appliances one after another. The turn-table ladders began picking them off the upper floors, while the escape, Dewhurst, hook and first floor ladders were used to all the lower windows. Within the first few minutes more than twenty-five people were brought to safety. The crews were magnificent, every man putting his all into the job at hand.

The fire was at the rear of the building and spreading rapidly. JP made Pumps Twelve, Turn-table Ladders Three as the fire began punching its way right up through the roof. After two hours of back-breaking toil, the fire was contained and then extinguished. Not one life was lost.

A couple of years later the Red Watch had the chance to put what they practise so often in the hopes of never having to use it, to the test. The call was to a fire in the Plough pub, Museum Street, just south of the British Museum. The landlord and his wife were laughing, joking and drinking with a few friends in the pub. It was long after closing time and they had all had more than enough. Above them their three children were asleep. In a nearby room on the top floor smoke was already drifting along the corridor towards their bedroom. The alarm was raised by the children's

Smoke curls out of a builder's temporary structure as if Aladdin's genie were about to appear. A large proportion of the structure is already alight, filling the smoke-laden sky with glorious colour.

dog, who, having barked and barked in the street but been ignored by the now drunken landlord and his wife, ran around the corner to the post office building, barking and whining until a postman came out to see what all the fuss was about. The dog howled, dashing back and forth along the road until the postman realized that the dog wanted him to follow.

By now the smoke was oily black and had crept like a snake under the door of the children's bedroom. The elder boy, aged ten, awoke with a start. There was now so much smoke he could hardly see his little brother and sister across the room. Without thinking he ran for the door, shouting at the others to wake up. As he opened the door, the fire, hungry for oxygen, leapt along the corridor towards him and charged into the children's room. In the distance they could already hear the distinctive clanging bells of the fire engines, as well as the blast of their two-tone horns. 'There's only one thing for it. Out the window, come on, look sharp,' said the elder boy. His brother was seven and his sister only five. They were so scared not one sound emitted from their throats. One after another they climbed out on to the parapet below their window. It was only six inches wide and they were on the fourth floor.

The pump escape, driven by Wally Slade, pulled up ten yards beyond the fire: perfect positioning to allow the crew to 'slip the escape'. Ted Temple was riding the ladders with Ian Macey:

Slipping the fifty-foot escape.

As we turned the corner we could see a lot of flame showing at the upper windows. The escape had already pulled up in front of us and it was being slipped. Glass came crashing down to the ground all around us. I glanced up and through the smoke I saw the three children. Flames were belching out of the windows either side of them, and in the wind were beginning to blow towards them. They made no sound, but just stared out in front of them, their backs pressed hard against the bricks. I ran up the escape and brought the first one down. As soon as I got her to the ground I went up for the little boy. A covering jet from Wally was working now, pushing the flames back in through the window. When I got back down with the second one, I collapsed on the ground completely knackered and puked my guts out, but God it was worth it! John Arrenberg, from Euston Fire Station, got the eldest brother down. Meanwhile the pub landlord and his wife were still downstairs, drinking away in the back, oblivious of what was happening.

On the very first day of the national firemen's strike, Monday 14 November 1977, a serious fire broke out in the old Charing Cross Hospital men's hostel. Soldiers manning Green Goddess civil defence fire engines fought this blaze under the direction of non-striking senior fire officers. A few weeks later, the strike growing more bitter and the opposing positions more entrenched, the 'red devil' came and stole away the lives of two small children from their home in Greek Street, less than one hundred yards from the fire station, the striking men and their equipment lying idle. The firemen on picket duty at the entrance to the station saw the Green Goddess fire engines clang their way along Shaftesbury Avenue and into Greek Street, but they were not given any details of the fire or told that children were trapped above the flames. Roy Baldwin, in his capacity as a senior officer, was not on strike and proceeded to the incident in his own car:

When I arrived, it was going well on two floors. At the same time a breathing apparatus team from the Royal Navy arrived. The soldiers were doing their best, but they had neither the equipment, training nor experience to cope with the fire. The Royal Air Force team were more highly skilled and began search and rescue operations. They appeared minutes later out of the smoke with three children. One we managed to revive, but two children had already lost their lives.

Mr Baldwin walked back down Greek Street where he met a group of the firemen who had by now come to the corner of the street to see what was going on. With grimy, tear-filled eyes he told them of the children. They

Top left: A four pump fire in a brothel in Old Compton Street. A dissatisfied client has petrol-bombed the staircase at the rear of the building. Just as the crew are about to enter the front door, a flash-over (backdraft) occurs. The burst of intense heat forces them to take cover on the ground. A few seconds later, jet full on, they enter the building to search for people who are reported trapped above.

Top right: Another clip joint is torched by a customer, this time in Rupert Street: are the girls still down in the basement? The front door is kicked open to reveal a blazing corridor.

Bottom left: Once the flames in the corridor are extinguished, Firemen Rodney Cordell and Bob Moulton discover that the fire is considerably more serious than it appeared from the street. Fire has already burned through the wooden steps of the staircase at the back of the building. Temporary Sub-Officer Roger Kendall passes a jet to Bob Moulton as he orders his driver to send a priority message to control: 'Make Pumps Four, Persons Reported.'

Bottom right: At the back of the building in Rupert Street, the fire roars up the staircase and bursts out like a volcano on the top floor. At the front of the building, no fire and very little smoke is showing, yet at the back the fire is so intense that Roger Kendall makes Pumps Six.

said, 'If only we'd known . . . God, why didn't somebody call us? Why didn't somebody tell us there were children who needed rescuing?'

After nine bitter weeks settlement was reached, pay was increased, working conditions were changed to meet union demands and the number of hours worked each week were reduced. But the national firemen's strike had been a catalogue of tragedy. Many people lost their lives in fires or lost homes which under normal circumstances would have been saved. The men of the London Fire Brigade had voted overwhelmingly against the strike, but the national results forced strike action. Men's hearts were broken when it came. One senior officer who loved the London Fire Brigade above all else in his life had a heart attack on the eve of the strike: a measure of the anguish the dispute caused. Dozens of other firemen resigned in disgust. It is a black cloud in a golden sky and one which the firemen who went through it prefer to forget.

Six White Watch firemen were nearly killed in a flashover during a fire on the *Old Caledonian*: a large floating public house moored beside Victoria Embankment. Knocker White called it the most frightening fire of his whole career:

It took us some time to get access as we couldn't get the key to unlock the main door which gave access to the lower decks. At last the door opened and me, Steve Holland, Steve Pearce, Peter McCarlie, Clive Wells and Raymond Taylor all edged forward in the black smoke, when all of a sudden, from being in total darkness, it was just as if the sun had come out two yards in front of our faces. The most brilliant blinding glow flashed over us. Poor 'Pixie' [Steve Pearce] was hit by the main force of the blast. The skin on his hands was turned inside out and he was just screaming and screaming in terrible pain. Steve Holland had dived the other way and his hands were even worse. They had lost so much skin they were just dripping in blood. It frightened the life out of me. I don't know how I got out untouched. Peter McCarlie kept low as well. The cone of his helmet was burned off. If he'd been six inches higher the blast would have taken his head off. Had we got into the boat sooner we would have been down the stairs into the main cabin and all six of us would have had it.

Two more rescues from fire are worthy of note, and both involved Station Officer 'Turk' Manning. The first occurred at teatime on 2 February 1983, in the Old Charing Cross Hospital in Agar Street, which over the years had already witnessed a number of serious fires, both when it was still a hospital and afterwards, when the building was used as a hostel for homeless men. The building housed 227 men of all ages and backgrounds.

A24

FIRE
FIRE

Some of them were ex-mental hospital patients, some were alcoholics and others had just fallen on hard times.

At precisely four o'clock the staff of the hostel held a fire drill. All the occupants were successfully evacuated, and within six minutes a roll call indicated that every person was accounted for. One of the residents, his pyromaniac tendencies presumably sparked by the drill, went back into the building, made his way into the basement and immediately lit a fire. Five minutes later someone on the ground floor noticed thick smoke emanating from the basement and broke a glass fire alarm point. Many of the residents ignored the alarm bells ringing as they had only just had the fire drill and assumed it was a false alarm. Luckily the staff dialled '999' as a precaution.

In three minutes the fire spread from one room to involve a large proportion of the basement, which was vast. Clouds of choking smoke filled the entire building, cutting off stairways and trapping dozens of men in their rooms. When Turk Manning and his Green Watch crew pulled up two minutes later, they were confronted with a building of six floors, 150 feet square, with men screaming for rescue on the roof and at windows on the third and fourth floors. Crowds of people lined the pavements, pointing and shouting, and dozens of other residents spewed like lemmings out of the ground floor exits of the building. One man was already climbing down a builder's ladder rushed to the scene by some enterprising locals.

The turn-table ladder immediately whirred into action, its crew successfully rescuing nine men from various windows, now belching black smoke. Turk made Pumps Six, Persons Reported, while his two very experienced Leading Firemen, Ron Blaber and Ian Southby, yelled like sergeant majors at the crews battling to get down into the scalding basement, the old Soho fire command first started by the legendary Mr Shawyer, 'Get in there!' The crews on the jet took a lot of punishment but eventually won the fight. One man was found dead in the basement, face down in the water, overcome by smoke and then submerged in the torrent of the fire-fighter's jets. Another man was rescued from the third floor, but he too was dead, despite strenuous efforts to resuscitate him.

At a much smaller incident Turk Manning nearly lost his own life. The fire involved a small quantity of burning plastic in offices above the department store D. H. Evans in Oxford Street. The fire crews wore breathing apparatus to tackle the fire as standard. Once the fire was extinguished Turk, not wearing BA, proceeded towards the incident, pushed open a large swing door and entered a corridor near to the fire which looked virtually smoke-free: a wispy haze was just visible. After just three steps forward in the innocuous-looking smoke, he collapsed: 'I suddenly felt dizzy, then incredibly sleepy. I stopped breathing and just slipped gently to the ground.' By chance Fireman Barry Humphries came into the corridor at the same moment and, seeing Turk unconscious, immediately began mouth to mouth resuscitation and massaged his chest until his heart started again. After about forty-five seconds Turk started to breathe again and although he did not immediately regain consciousness his life was saved.

The smell of Chinese food fills the air. Sunday morning in Chinatown. The traffic. The noise. The ever-present smell so distinctive of Gerrard Street. To a stranger the exotic smells conjure up appetites and romantic visions of the orient. To the men of Soho Fire Station the constant stink of frying wonton is sickly and pervasive.

Chinese immigrants have lived in England since the late eighteenth century. They were all sailors and those that settled chose the ports of Liverpool or London. By 1900 their population was still tiny: less than six hundred in the entire country, almost all men, whose shops and cafes catered mainly for the transient Chinese population of seamen. By the outbreak of the First World War in 1914 some thirty Chinese businesses were established in the Limehouse area of London, which became widely known as Chinatown. However, during the Second World War, the Limehouse district of London was flattened and, in the fifties, new union rules made it increasingly hard for non-British seamen to get work:

> Suddenly the British Chinese community . . . was faced with a double problem: finding new sources of income and finding a new area to live . . . Gerrard Street in the late fifties was one long line of shabby brothels, seedy night-clubs and faded restaurants . . . short leases could be picked up for next to nothing in the whole area between Shaftesbury Avenue and the back of Leicester Square . . . Chinese entrepreneurs began to acquire premises in Gerrard Street and Lisle Street in order to set up restaurants of their own. At the same time, thousands of agricultural workers from Hong Kong's New Territories, forced out of their traditional occupations by changes in the world rice markets, began to arrive in England looking for work.[1]

By the early seventies Gerrard Street had become the centre of the new Chinese community, and it has continued to grow ever since. In addition, men from Singapore, Malaysia and Vietnam arrived in their hundreds, soon to be followed by their families, all eager to set up businesses and move in on the Chinese monopoly.

The night shift of Sunday 18 July 1982 was proving to be another

[1] Judith Summers *Soho* (Bloomsbury, 1989)

Three of Soho's crew, Wally Slade, Micky Seals and Ted Temple, inch forwards as they prepare to enter the ground floor of Mister Byrite clothes store in Oxford Street, 11 January 1982. Their jet has knocked the flames back into the shop, but the orange glow from within warns them of the intensity of the fire.

record breaker for the Blue Watch: Station Officer Tony Wilmott, held in great respect by his men and probably the most cool-headed fire officer in the brigade, was leading his crews to defeat a six pump fire in Charing Cross Road. During the height of this fire, his driver informed him that another four pump fire was in progress in Kemble Street, near the Aldwych, also on Soho's ground. He cursed his luck that there should be two 'good jobs' going on his ground at the same time.

Once the fire in Charing Cross Road was surrounded, Soho's pump escape crew restowed their gear and headed back to the station. Sub-Officer Bill Neal, known as 'Pug' (after the men of Soho started a rumour that the popular children's character, Captain Pugwash, was based on him), ordered the escape's crew to change their breathing apparatus cylinders, have a wash and a cup of tea, and prepare to go back 'on the run'. As they were doing this, the 'running call' fire bells rang out their distinctive repeating alarm. The dutyman opened an appliance bay door and met a Chinaman in a state of great alarm and agitation.

'Fire, fire . . . come quick, come quick,' he panted.

The dutyman was firm and assertive: 'OK, mate, calm down a minute. Tell me where the fire is.'

'Fire, fire! Very bad, please come quick,' he replied, his voice rising in anger and exasperation at the fireman's apparent desire to hold some kind

Three seconds later, a flash-over (backdraft) explodes out into the street. The force of the fireball is so intense that it demolishes the front of the shop, which crashes down on to the pavement where the crew were kneeling seconds before.

of conversation with him. Now the fireman's voice rose higher in pitch too: 'What street, what street?'

'Hah!' The Chinaman's eyes sparkled as the light went on. 'Gerrard Street!'

'Pug' reached for the RT as the appliance swung out of the station: 'M2FN from Alpha 2.4.1 priority . . . running call to fire, Gerrard Street. Soho's pump escape attending, Sub-Officer in charge.' The appliance surged forward along Shaftesbury Avenue, spun left into Gerrard Place and right into Gerrard Street. Halfway down thick clouds of black smoke billowed out of the basement and ground floor of number 39. 'Pug' made Pumps Four, Persons Reported, and the escape crew very quickly got a jet to work and began to make their way into the blazing basement.

Once the fire was out the crew found two bodies in the basement quite near to the exit. Both were obviously dead. Then two more were discovered. Then two more again. And then a seventh. This last man showed some signs of life and was removed to the open air. Witnesses nearby reported hearing an explosion and seeing a roar of flames from the basement steps. The basement was one of several illegal drinking clubs in Chinatown. The police had been called to a gang fight several hours earlier outside number 39 and began an immediate mass murder investigation.

By chance a doctor, Kenneth Hines, was riding as an emergency incident doctor with one of the London Ambulance Service crews who were dispatched to the fire. Suspecting foul play, the police officers ordered the firemen not to touch the remaining six bodies in the basement. However, as a fireman's first duty is to life not the law, the senior fire officer present made a compromise: Dr Hines, who was fully trained in the use of breathing apparatus, entered the blackened basement accompanied by two fire officers, crawled through the still steaming, charred rubble, located the bodies still shrouded in thick smoke and began examining them. Had the officers and doctor not been wearing breathing apparatus they would have smelt the gas! Gallons and gallons of gas, pouring into the basement from a ruptured main. Silent, deadly and undetected. In amongst the soaking ruins a tiny hot spot glowed an evil red, like a stolen jewel winking in the darkness. With each second that passed, more and more gas built up in the basement, taking the atmosphere closer and closer to the critical explosive mixture, like a time bomb ticking. In the gloom the doctor knelt beside the body of the sixth man. He shook his head, indicating that he too was dead, and led the team back out and up to the street.

The BA control point (where the tallies of each BA wearer are slotted on a display board and monitored by a fire-fighter to keep an external record of exactly how many fire-fighters are committed to a fire whilst wearing BA, including a record of the time they 'started up' their sets, the amount of air in their cylinders and the time they are due back out of the fire) was set up on the pavement outside number 37. Dr Hines glanced at his watch: 0245 – almost exactly an hour after the initial call. He wiped the sweat from his eyes, blinking away the visions of hell he had just seen. In the basement a draft of air nurtured the glowing eye of fire, which rekindled. The flame grew all at once from the size of a match burning to a roaring, spitting volcano: the gas exploded, shooting flames through the ceiling of the basement and up through the centre of the building in a great whirling wind. The large plate glass window across the shop frontage on the ground floor burst into the street in a thousand flying pieces. Two police officers, a chief superintendent and a constable, were cut down in a hail of razor-sharp fragments. A fireman nearby was also hit in the back. Miraculously no-one else was hurt and by chance no firemen were in the building. Dr Hines rushed forward and, assisted by the ambulance crews, carried out on the spot field surgery on the severed arteries of the officers. In doing so he undoubtedly saved their lives.

The London Fire Brigade in general has a wonderful *esprit de corps* and at Soho this is particularly pronounced. Ernie Allday, ex-Soho Sub-Officer and now retired Assistant Chief Officer, has many memories:

In the fifties we had had several serious fires in the supply tunnels under London, including a very protracted incident in Queen Victoria Street (Cannon Street's ground). It was therefore important that the officers and firemen of Soho knew their way around the complicated maze of tunnels that runs from Leicester Square up Charing Cross Road and another down Shaftesbury Avenue as far as Piccadilly Circus. It carries all the supplies: water, gas, electricity and in those days even hydraulic mains. Inspection covers gave access from central traffic islands at various points along Shaftesbury Avenue. There is a network of fabulously complex tunnels, some running beneath the underground tunnels of the Piccadilly line. A shocking labyrinth of deep excavations, which we used to explore to familiarize ourselves with the layout in case there was a fire down there.

One evening I was down one of these tunnels with Leading Fireman Freddy Lane, known as 'Shady', who was a good fireman with a lovely sense of humour. We had been down as far as the Circus and were walking back towards the fire station. Eventually we came to an exit ladder halfway up Shaftesbury Avenue. Freddy climbed up first and as he neared the metal grate he saw that a man was standing directly above, waiting to cross the road. Quick as a flash Freddy tapped on the metal with the keys. The man was most confused, and glanced about, embarrassed that he could not explain the strange noise. Freddy gave the metal just inches from his feet another sharp rap. The man looked down and almost jumped out of his shoes in surprise! Freddy's eyes were wide and white like a minstrel as he beckoned him conspiratorially to come closer. The good man got down on his hands and knees and pressed his ear to the grate. In a hoarse stage whisper Freddy questioned him ever so seriously, 'Is the war over yet?'

'Yes,' he replied with great concern on his face.

'Well, then, do us a favour and let us out,' said Freddy, as he handed the keys ceremonially up into the chaos of busy theatreland.

We also used to have numerous calls to chimney fires. And in the big houses, the clubs especially, the old boys would pop themselves into a big leather armchair after a good meal and throw an extra shovel of coal on to an already huge roaring fire. Some of the buildings had support beams through the chimneys. Furthermore, the chimneys were never straight up, and it was often a hell of a job to find the right chimney as they would snake and ladder behind the scenes, forming a warren of interconnecting shafts. The old boys didn't want to be disturbed. We often used to have to go up on to the roof and pour water down from the top.

One time at the Devonshire Club we had to tear down the wall to get at the fire which had spread along the beam. The old boys would not move, they all sat there puffing on their cigars, shrouded in clouds of

18 July 1982: seven men are already dead – murdered – inside this building in Gerrard Street. One of Soho's crew holds his helmet between his legs as he dons his breathing-apparatus face-mask. A Station Officer waits for him on the basement stairs. The fully charged jet of water is ready for use. The ground is littered with broken glass from windows on the upper floors. The large ground floor window was later blown out in a gas explosion which occurred once the main fire had been extinguished.

The burning *Old Caledonian*, 27 April 1980, in which five firemen were injured in a flash-over. She dwarfs the firemen standing on the access jetty between the ship and the embankment.

plaster and dust. They looked like museum pieces by the end of it. The funniest mistake of all which was not uncommon was to pour water down the wrong chimney – smoke would often come out of two or three – which would cause a fall of soot that covered everyone in the room below.

At Soho you had to be so wary, you could never take anything for granted. You have to be so careful which door you went through. One day we had a job in Dean Street, a house on a corner. We broke in through a side door and made our way up the smoky staircase. On each floor there were two doors. They were all padlocked heavily and I assumed that they were workshops. To find the fire it was just a question of choosing the right one. I felt both doors but they were cool.

'Try this one, Eric.' No luck. 'Try this one, then.' Again no luck. Bash, bash, bash went the sledgehammer: those locks kept coming off but still we couldn't find the fire. Floor after floor, the locks crashed to the ground. On about the fourth floor we opened yet another door and in I went. To my surprise I found myself in a cupboard full of clothes. I pushed the door of the cupboard open and stepped out into a bedroom. The man and woman sat bolt upright in bed, confronted by a fireman with a large axe. He was a very well-known gentleman, but I shan't tell you his name, as that wouldn't be right.

Another time we were called to Charlotte Street. The lady on the fourth floor said she had seen smoke coming out of the tailors' workshop on the floor below her. As so often in Soho nothing was showing at the front, but smoke was coming out of a window at the back. We broke in and found that an iron had been left switched on and had burnt through the wooden bench it was on, leaving the clear shape of an iron as if it had been cut out of cardboard. The iron had then fallen on to some clothing below, which was burning well. It was a good hose-reel job.

Just as we were about to leave, one of the firemen glanced out of the window and noticed smoke drifting past outside. He stuck his head out, and gasped in amazement. 'Guvnor, I know you're not going to believe this, but there's smoke issuing from the first floor below us!' We all dashed downstairs and broke in there. The chances of what we found must be thousands to one. The occupier was another clothing manufacturer of a completely separate company, yet we found another iron had burned through another wooden bench in exactly the same manner.

Terry Spindlow, who served at Soho from 1958 to 1974, can also hold younger listeners enthralled:

The thing about Soho's ground that makes it so special is the unique combination of architecture and multiple occupancy. Many a time we would pull up at a job in a typical Soho street like Dean Street. There would be smoke issuing from the second floor at the front, so we'd slip the escape to that and make our entry there. At the same time other crews would be making an entry on the ground floor. But we'd often find that breaking down one door would lead you into a warren of dead ends, each one requiring another door to be smashed, and all the time the fire is raging away untamed because we haven't actually been able to find it. I remember one day Ferdie Hurcombe was in charge and we had to break seventeen different doors down before we actually located the fire. To an outsider this might sound like incompetence and people sometimes used to accuse us of being a little light-fingered on the sledgehammer trigger, but it was not the case. Many of the buildings are so complicated inside that you truly cannot ever judge a Soho building by its cover.

When I went to Soho, John Shawyer, brother of the equally famous Alfie, was Guvnor of the Red Watch and Shady Lane Guvnor of the Blue Watch. I remember a visit to the British Museum Library in the early sixties when Ferdie Hurcombe was Station Officer of the Reds. We looked at some old log books from the station and found that Ferdie's father had been in the watchroom at Soho in 1930.

In 1958 there were just the Red and Blue Watches, just as it had been since before the Second World War. We did a week of days and a week of nights: that was a sixty-hour week, including a twenty-three-hour Sunday shift, which was murder when you were busy. In the watchroom, which was still continuously manned, there was a bank of some sixty-five fire telephones providing direct lines to all the important buildings, museums, hotels and hospitals on the ground. Each line used to be tested twice daily. It was a hell of a job for the watchroom man, taking several hours to complete, always starting with St James's Palace first at 0615 each morning. In addition we had thirty-five automatic fire alarms from various department stores and a couple of large office blocks. These had to be tested every day as well.

Soho was unique . . . I miss it terribly. In that one tiny place we had every type of risk you can imagine. We even had a haystack fire in Trafalgar Square one Christmas. The hay had been brought in for a nativity scene!

Another thing that makes being a fireman in Soho so special is the grand cast of local characters who populate the area. There was one old tramp who always wore suits. He obviously had a certain knack of getting them as every month or so he would arrive outside the station, wait until the pavements were crowded and then he'd strip off his old suit, which by that time was soiled and rotten, and pompously put on his new one. Then he'd wave to the firemen, say, 'All right boys' in a 'Hello sailor' voice, blow them a kiss and mince off in his dapper new suit, though his hole-filled, coffee-stained shoes were always a bit of a give-away.

Another regular visitor to the front of house unofficial pavement theatre was an ex-Gurkha who, falling on hard times, finished up as a tramp living in and around Shaftesbury Avenue. Never without a bottle of wine, and always with a grin to match his high spirits, he would perform his own version of 'Land of Hope and Glory', shout 'God Save Her Graciousness', drink the sovereign's health and shuffle off into the night.

The station today still has some wonderful characters behind the big red doors. Burt Wilson has been on the Blue Watch for as long as anyone can remember and has also been the watch mess manager. Tens of thousands of meals have been created by his culinary hands and none more memorable than the night of the 'botulism pie' saga, when half the watch went down with food poisoning. Burt had made a pie with black mince that wasn't quite cooked enough to kill all the germs. 'The Guvnor was at home digging a hole in the garden when he fell straight in, struck down like lightning from the hand of God he was, but it was Burt's pie that had done it!' recalled 'Pug' Neal with glee. Once the men gathered around the table started into Burt the stories began to flow:

Another time he served up some corned beef fritters made from War Department meat that was left over from the war: some stuff that Rommel had left behind in the desert. I came up a bit late from the office, opened the dormer doors to get my food out from the hot-plate and there it was piled high with fifteen fritters. The old cry went up, 'Don't be last' as I had to eat my way through a pile of beefburgers that were more like ice-hockey pucks! And then there was the time he got some sausages that were rancid: you could taste that they were bad, but Burt said they were just spicy. The next night he produced a job lot of fishfingers that were filled with sawdust. They cost him about ninepence a thousand . . .

And so they went on, story after story, laughing and laughing until their sides almost split.

Theatre fires are very rare now, thanks particularly to the strenuous efforts of Mr Eyre Massey Shaw, Chief Fire Officer of the Victorian Brigade, following the series of terrible fires that destroyed most of Soho's theatres (see Chapter 2). All theatres have their own duty firemen, themselves retired London fire-fighters, who patrol the theatre day and night. Most modern incidents in theatres occur during building work. Station Officer John Peen was called to a 'Fly curtain alight' at the Aldwych Theatre. Both the front of house and the side of the stage were in darkness:

I was standing in the wings and I could see that the curtain was smouldering, caused by a spotlight touching it. Suddenly, as I was walking around in the gloom to get a better look, I realized I was standing next to a coffin. Hamlet was in full swing, the actors were in the wings on the other side and I realized I was partaking in my West End stage debut, in full view of the audience! They loved it.

Firemen 'Speedy' Close (left) and Clive Robinson emerging from the Astral Cinema, Brewer Street, following a serious fire in the basement.

Dave Smith recalled a time when in the course of a smoky job he turned to the fireman next to him, talking away, Cybermen-like in their breathing-apparatus masks:

After a couple of seconds I realized he was not replying to any of the questions I had asked him. I moved closer. He moved closer too. I moved my right arm towards him. He moved his left arm to meet it: I nearly jumped out of my skin. It was me. In a mirror in this corridor. Spooked the life out of me it did!

No story of the notable fires of Soho would be complete without at least a passing reference to the Sundowner Club. The initial call was to a fire in a night club which occupied a disused cinema on the corner of Denmark Street and Charing Cross Road. Station Officer John Peen was in charge

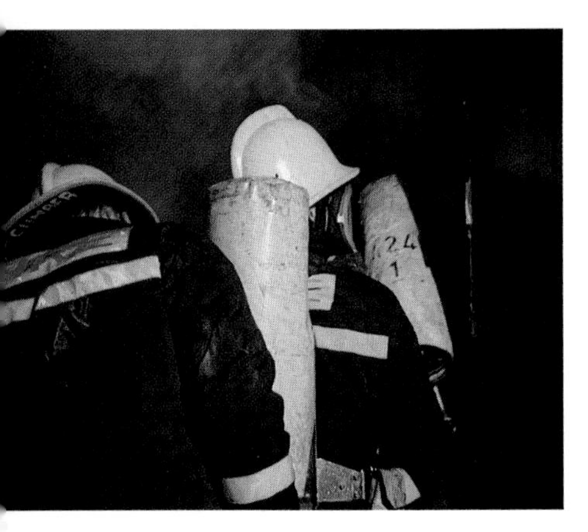

A breathing-apparatus crew, identifiable as Soho men by the markings on their cylinder covers, enter the rear of a building in Cockspur Street, off Trafalgar Square, where an eight pump fire is in progress in the basement.

'We were following our instincts as we went into the job,' says Fireman Dave Smith. 'It was red hot going up the main stairs and everything about it suggested the fire was in front of us. Suddenly there was a cracking great whoosh, and this flame shot up past us from the basement. I turned on my back and slid the full length of the staircase, using my cylinder like a surf-board. I didn't hurt myself because of the adrenalin pumping through my body. No matter how many years you've had in the job, you always make a muck up once or twice. It's something to do with the unpredictability of fire.'

of the White Watch that night. There had been several small fires on the premises in previous months and they knew the building well: what they did not know was that the entire cast of the musical *Tommy* had taken over the club for a first night celebration party. 'All the stars and nutters were there, and the drink was flowing like it was going out of fashion. As rock music burst people's eardrums, we forced our way through the wall-to-wall crumpet, trying to locate the source of the smoke that hung in the air.'

It turned out that one of the many live cabaret performers drafted in to make the party really swing, a fire-eater, had set light to the snake charmer's basket. In all the excitement the basket had been knocked over and three pythons were loose in the crowd. Moments after the firemen arrived the disc jockey put on 'Come on, baby, light my fire' as a special. On the stage across the room, a trick shooter raised his .22 pistol and prepared to perform his act, aiming the gun at a target behind him by pointing it blind over his shoulder. As his finger rested on the trigger he caught sight of one of the pythons making directly for his ankle. Bang! The gun went off and shot the DJ in the neck; he fell bleeding to the ground. The crowd thought it was a great act and cheered and shouted their approval. When the DJ stayed down on the ground, several of Soho's firemen rushed forwards to administer first aid. The police swamped the place and said that nobody could leave, as they were beginning an attempted murder investigation, and things went from bad to worse. The Sub-Officer was assaulted by a big German, so another fireman gave him a crotch clutch-hold and lifted him up, but the club had a very low ceiling, so that left him 'sparko' on the floor. Chaos ensued with enthusiastic girls dragging reluctant, fully rigged firemen on to the dance floor. The club was filled with press photographers and reporters who rushed away to make the most of the story . . . Walk into a room full of showbiz celebrities or firemen, even after all these years, whisper 'The Sundowner', and anyone who was there that night will either wink, smile or turn white.

Each year Soho's fire-fighters receive more and more calls for help. While the most important aspect of their work is still the prevention, containment and extinguishing of fires, a higher and higher proportion of the calls to which they respond are not to fires at all. These non-fire emergency calls are collectively known as special services.

Special Services

Special service calls cover almost any mishap you can imagine, from endless requests to assist careless residents locked out of their flats to the time-consuming and often dangerous work of dealing with spillages of hazardous chemicals, gas leaks, floods, burst water mains, bomb explosions, collapsed buildings, accidents with mechanical equipment, sewage workers overcome by fumes and collapsed in tunnels deep under ground, bits of buildings damaged in storms left hanging precariously over pavements below, even people with rings stuck on their fingers . . .

The most common of the special service calls is to people who have become shut in lifts. A high proportion of the buildings in Soho have lifts, ranging from gloriously gargoyled Victorian cast-iron monstrosities in the gentlemen's clubs to ultra-modern high-tech glass elevators in swish Covent Garden advertising company offices. Many of the grand hotels in the area have splendid modern lifts for their guests to use, but behind the scenes, below stairs, there exists a hidden world of rickety old staff and service elevators.

Power cuts have on occasion caused havoc: in 1990 there was a major power cut in the West End of London following an explosion in an electrical junction box. Hundreds of people instantly became stuck in lifts all across Soho's ground, in hotels, hospitals, shops, offices, clubs, cinemas, theatres and underground stations. In a matter of minutes Command and Mobilising Centre – the London Fire Brigade's new ultra-modern computerized but extremely unergonomic '999' command centre – received hundreds of calls to people shut in lifts, and machines from fire stations as far away as Chelsea and Lambeth were summoned to assist.

Usually, however, the culprit is faulty machinery causing the lifts to stop between floors or else the doors simply fail to open. The fire crews proceed to the lift machinery room, switch off the electrical supply and operate manual winding gear by means of a crank handle. Once the lift is lowered or raised to the nearest floor, a lift key is inserted into a small hole above the doors, enabling them to be opened manually.

Late one evening at Covent Garden underground station a man was saying goodnight to his girlfriend. As they parted he hesitated for a split second before stepping back into what he thought was the lift behind him. At that moment the inner lift door and outer shaft doors closed simultaneously, trapping him between the two sets of closed doors. The lift dropped away, leaving the poor man marooned on the tiny ledge with a seventy-foot vertical drop beneath him and nothing to hold on to on the inside of the wooden door. The external doors were made of wood and contained square glass panels at head height. His girlfriend saw his face through the glass, white as a sheet, his eyes imploring her for help. She alerted the station staff and the fire brigade was summoned. On arrival – which, although it only took three minutes, must have seemed an eternity to both the man and his girlfriend – the firemen broke the glass of the panels on either side of him, ran a line around his shoulders, and then, once he was secure, opened the outer doors by hand and delivered him into the arms of his beloved. They went home together that night!

One of the things that makes Soho such an unusual fire ground is that, with the exception of buildings constructed in the last ten years, any building you go into is no longer being used for the purpose for which it was originally designed. This means that there will always be an architectural or structural complication. For example, lift motor rooms (where you isolate the power and where the hand-winding gear is operated) are normally located either above or below the lift shaft. But on Soho's ground there are buildings where the lift motor room might be in a nearby building or at the most extreme in another street, connected via cables running usually at sub-basement level in what had once been all one building, but which had subsequently been divided and converted. Firemen arrive in the middle of the night: there's no-one to tell them about the layout of the building and they are forced to play 'hunt the missing motor room' at 0330 in the morning.

In the early fifties, Fireman George Phillips and his colleagues were called to Golden Square, at that time still a major centre of the 'rag trade'. The building in question was typical in that it had a pavement light which opened to reveal a small hydraulic lift, used to take material deliveries from street level to the basement. On this morning the cleaning lady must have noticed a cobweb on the underside of the pavement light as she was dusting in the basement. As she moved forward on to the lift she knocked the lever into the 'on' position. Silent and deadly as a killer shark, the lift very gently began to rise. So smooth was the ride that she did not even notice that she was getting closer and closer to the pavement level, until it was too late. An unfortunate passer-by happened to glance down and to his horror saw a body squashed tight against the underside of the straining pavement light; it was he who summoned the brigade. The rescue was not difficult: the men simply switched the lever into reverse, but the cleaning lady's body was covered in a macabre patchwork indentation. Occasionally people manage to get themselves caught in the actual machinery of lifts. Sub-Officer Terry Spindlow was called to 'Man trapped in lift, at the post office, Broadwick Street'. When he and his crew arrived he saw 'the upper torso of a man sticking out from under the lip of the goods lift, at pavement level. There was a clear sign on it that read "Goods Only, No Passengers". As I approached the man I noticed he was horribly still. I thought he was foreign because he was so dark-skinned. In fact he was a white bloke, but his skin had gone dark because he'd been crushed in half. Gone black through suffocation. Horrible.'

Sub-Officer Bill Neal remembers two other special services:

Temporary Station Officer Dick Haigh shouts instructions to his crew, who are sweating away in the lift machinery room three floors above. They are winding the lift up by hand to release a family of three, stuck for four hours before the alarm was raised.

We once had a bloke brought to the station in an ambulance with his arm caught in the back of a reclining office executive's chair. We had a vice on the station in those days, so we strapped him into it and then sawed him out. Another time a bus conductor got his hands glued to the rail at the bottom of the stairs of his double-decker bus. Some stupid kids had put Superglue on the metal rail and he was stuck fast. We had to cut the rail off on either side of his hand, and he was more worried about the damage to the bus than his hands.

'Look, mate,' I told him, 'we've got to cut it off to get you to hospital.'

'Ah no,' he said with a long sigh, 'the inspector won't like that.'

When Roy Baldwin was Assistant Divisional Officer, A Division, Soho had another unusual rescue to perform:

The call was made by the police at a public toilet in Shaftesbury Avenue, where a woman was trapped in a toilet. I attended because a person was trapped, but on the way I thought to myself, 'She's probably just stuck behind a locked door.' However, on arrival, we found that the poor woman was actually stuck on the loo seat. It was an old wooden seat and she must have been in a hurry and sat down in a rush. The seat had come over the shape of her and jammed solid. We had to cut the seat away with a saw, but there wasn't much room and it got awfully hot and sweaty. The firemen made the usual polite conversation with the lady while they were cutting her out, but under the circumstances they had to take extra care to avoid their lavatorial humour.

People sometimes get themselves actually trapped in machinery. Over the years Soho's men, often assisted by the crew of Euston's Emergency Rescue Tender (a purpose-built mobile workshop carrying a highly specialized selection of rescue equipment, heavy lifting gear, cutting and hydraulic spreading equipment, a complete collection of tools, winches, clamps, saws, jigs, in fact every conceivable gadget) have rescued people from the most surprising situations. These include both humans and dogs who have become caught in moving staircases when their clothes or limbs have been trapped and then dragged into the machinery. Wherever humans work with machines they seem, sooner or later, to find some way of getting themselves stuck in them: refuse collection lorries, meat-mincing machines, printing presses, material stretchers, even washing machines!

The antics of stag-night revellers lead to requests for rescue too: on several occasions drunken men have arrived on the doorstep of the station, their wrists raw red from wrestling with handcuffs. Occasionally they are cuffed to friends, and on one occasion, to a complete stranger! Release and relief comes quickly with the aid of cutting equipment or, more often, a simple fret-saw. Sometimes the unfortunate groom-to-be has been chained, usually naked, to railings, in which case the firemen are required to take their equipment to the scene of the crime, bringing even greater crowds to gawp and stare at a time when the luckless victim is usually wishing he were completely invisible, especially on a cold night.

Another regular special service is to people who are locked in or out of their dwellings: sulking children lock themselves behind bathroom doors, push bolts closed or shut latched doors unexpectedly. Soho's crews often have to comfort frantic mothers as they break open doors and windows. Sometimes office workers, staying behind late in the evenings, are locked in by security staff. More than once, very embarrassed-looking businessmen have been rescued along with their secretaries from West End offices.

Firemen have truly to be jacks of all trades. Included in their multi-talents must be expertise as cat-burglars. Windows or doors are only broken if there is no possible alternative. Many is the time that firemen have gained access to flats via apparently impossible means of entry: tiny third floor windows not much larger than cat flaps. Such are their skills that it is not unknown for cocky burglars to call the fire brigade to assist them to gain entry to flats – as a result the police are now requested to attend lock-out incidents to verify the ownership of the caller.

Sub-Officer Steve Short attended one particular 'lock-out' which he will never forget:

The lady concerned was a sweet old dear of at least eighty. She came to the station saying she had locked herself out of her flat in Shaftesbury Avenue. She showed us to her block, where the front door was open. She led the way up the stairs, but it took ages for us to reach her flat because it was on the fifth floor and I thought to myself how awful it was that such an old lady had to live so high up a building. She showed us the door to the flat which was at the back of the building. There was no possible access with ladders from the rear, so I explained to her that we would have to break the door down. 'Oh yes, that's fine,' she said, smiling. One of the boys kicked in the door, which flew open with a great crack of splintering wood. The old lady walked into the flat, turned round to me and said, ever so casually, 'This isn't my flat, dear.'

My voice went all high like something from a Laurel and Hardy sketch – 'This isn't your flat?' – then I turned to the fireman who had just kicked in the door and repeated myself, as if he hadn't heard the first time, 'This isn't her flat.' I was simultaneously trying to lick the splintered wood back into place. At that moment I spotted a postcard on a side table in the corridor. Glancing at the name on the card, I noticed that it was the same surname as the old lady's.

Soho's turn-table ladder serves as an external staircase for an unfortunate secretary locked in her office after staying on to work late.

The mystery was solved: it had been her flat, ten years before. It was her son's place now and she lived in an old people's home in the country. Somehow she had found her way back to London and in her slightly confused state had assumed she had lost her keys.

'Knocker' White once had a call in the middle of the night to 'Woman locked in bathroom, Walkers Court, near Brewer Street':

When we arrived we rang the doorbell and to our surprise were met by a very voluptuous and very beautiful lady who looked in great shape. In fact she looked absolutely marvellous: she had no clothes on. Far from being locked in she was completely free from any constraints at all! She'd just felt like some 'company'. The Guvnor was furious with her and made us leave immediately. Well, he had no choice, did he?

The West End has more than its share of London's wrought iron railings. In the late eighties the bells went down, summoning a crew to 'Person stuck on railings, Temple Place, near the Embankment'. On arrival they found a local tramp hanging head down, at full stretch, his foot fully impaled on the railing spike. Apparently he had been attempting to climb over the locked gate, lost his balance as he reached the critical part of the exercise and brought his foot down full weight on the top of the spike. Somehow he managed to perch there on top of the railings. He howled in

'Knocker' White and Tim Alderman (left) on the balcony at the rear of the old Soho Fire Station, 1971.

agony and soon a group of passers-by had gathered around him to hold him in place until the rescue party arrived. (Whenever a special service call involving injury or accident is received at the London Fire Brigade control centre at Lambeth, the control operator will telephone London Ambulance Control and the Metropolitan Police. The same inter-services communication occurs if the police or ambulance service controls receive a '999' emergency call which they believe will require the attendance of the fire brigade.) It was quickly decided to cut the spike off the railings with it still sticking through the tramp's foot. He was whisked off to hospital where the spike was subsequently removed by surgeons.

Firemen are often avid readers and many of them scorn popular fiction in favour of philosophy, classic literature, history and psychology. Had Station Officer Turk Manning read Carl Jung's *Synchronicity* he would not have been so surprised by the extraordinary coincidence he experienced in the company of an Irish road driller. Turk recalled:

Soho's pump was called to rescue a workman from the roof of the newly constructed Household Cavalry barracks in Knightsbridge. The workman had been drilling in the street and had put his drill straight through the main power cable. The force of the shock had blown him quite literally out of his boots: he came down with a great crash on the roof of the barracks, drill and all, but somehow he survived! His boots were still smoking in the street when we arrived, just like something in a kids' cartoon.

Six years later, our pump was called to Wellington Street, near the junction with the Strand. The call was to 'Workman electrocuted by drilling into main electricity cable'. When we arrived we were confronted with this big hole, the drill sticking out of it, all sizzled and smoking, but no sign of the workman anywhere. I glanced disbelievingly at the buildings around me, remembering the previous time this had happened.

'All right, where is he?' I asked.

'We've taken him into the pub,' a civilian replied.

In the gloom of the pub I made over to where he was, laid out on a seat. He looked in a bad way, so I put my face close to his and whispered, 'What's your name, mate?'

He paused and then said in a wonderful Irish drawl, 'Finneas Fitzpatrick.'

Suddenly the unbelievable truth began to dawn on me, so when he'd calmed down a bit I said to him, 'This is the second time this has happened, isn't it?', at which point he recognized me. He just would not believe it and nor could I. That anyone could survive such a terrible shock to the system once is amazing enough, but twice!

Flooding calls are almost inevitable in Soho whenever there is a really heavy rain storm: there is probably not a single building in Soho that has no basement. Some buildings have numerous basements, including government buildings whose design and structures are top secret. (In the event of an emergency the fire brigade have complete right of access to any building with the exception of Crown property and any buildings with diplomatic status. If an emergency occurs in a government building the officer in charge of the fire appliances at the scene has to be invited to enter the premises by someone in authority.) Divisional Officer Roy Baldwin recalled a particularly difficult flooding:

It was during the Falklands War and at a time when the Fire Brigades' Union was making a fuss about the use of our lightweight portable pumps: there was some controversy over their safety: the user might get an electric shock if you put them in water because they did not have any adapters on them. As a result the pumps had been taken off the run.

Anyway, we had a flooding in part of the Admiralty in Whitehall. It was a very long way down: four floors down from the ground floor and then a lift down even further! Water was seeping through from a floodgate under the Thames. We entered a series of complex tunnels and finally came out into a street in the City, almost a mile away. The flooding was particularly dangerous as the top secret communications centre from which many of the messages were being sent to the South Atlantic was nearby. I realized very quickly that the only way we were going to be able to pump out the water was with the portable pumps, so I sent a priority message requesting four of them; but fire control came back with the response 'Answer no, they are off the run.'

Meanwhile the flooding was increasing and the electrics for this control centre were threatened. I got on to control by land-line [the fire brigade jargon for using a telephone rather than a radio telephone. Land-lines are used to send messages which are of particular sensitivity to control] and said, 'Get on to the Chief Officer and tell him I need them and now.'

We got them, and Soho's men did a magnificent job. Because of the secrecy of the location nothing was ever said, and even now I cannot reveal the exact location of the incident.

Soon after coming to Soho as Station Officer on the Green Watch, Chris Staynings found himself playing therapist and counsellor to a crestfallen cyclist on the top of a building in Steven Street. Calls involving people threatening to jump off buildings are collectively known as 'jumpers' and such calls are common on Soho's ground:

It was absolutely freezing that day. The man was dressed in a cycling suit and nothing else, with the wind howling over the top of the building. He chain-smoked the entire time. Poor guy, he was shaking and shivering, so like a fool I gave him my tunic and all I had on underneath was a tee-shirt. So he was shaking like a leaf, just slightly less than before, but now I was shaking as well. After about an hour of talking I moved closer towards him until there was only ten feet between us, but he panicked and for a moment I thought he was going to launch himself off. I was trying to think of things to do and say, but he wanted to do most of the talking.

The trouble was, he'd bought a mountain bike on hire purchase to get a job as a dispatch rider. The next day it had been nicked and he owed three hundred quid. So he had no job to pay for it and no insurance. So what does he do? Nick another one to keep his job . . . but he got caught and fined eighty quid by the magistrate. He was really desperate. The only good thing that had happened to him recently was that he had written several articles for a cycle magazine, where he had a friend who was the editor. As a special favour the editor lent him a brand new road bike which was on loan to the magazine. The deal was that he could use it to keep his job on condition that he wrote a review article on this new bike at the end of the trial period. But the bike was stolen from him the very first night. At that point he had decided to end it all by jumping off the nearest high building he found open. Once he had told me his story he seemed more relieved and after nearly two hours he relented and we took him down to the ground floor.

Experience has taught the men of Soho that if the person threatening to jump is in a very public place, the chances are against them actually jumping. One sunny day the Red Watch were called to a girl threatening to jump off Nelson's Column, Trafalgar Square. On arrival, as always by silent approach (that means driving to the emergency without sirens or flashing beacons so as to avoid the risk of frightening the person into jumping), Station Officer Colin Townsley and his crew were confronted with a young punk girl standing some twenty-five feet above the ground threatening to throw herself off, her pink and orange hair flying like a battle standard in the morning breeze. It was a glorious summer's day and the crowds were already gathering in their thousands; people in buses strained their necks to get a better view of the drama. Amongst the many driving past the scene were two men in an open top car. To everyone's shock, one of them shouted at the top of his voice, 'Jump, you fucker, jump!', echoing the infamous lines from a Peter Cook and Dudley Moore sketch. The punk girl looked quite surprised herself.

We looked at each other in disbelief and held our breath. Would she do it? She sat down on the edge, pushed her arms behind her and prepared to launch herself off. The now large crowd drew in its breath collectively like spectators at a fireworks display. Colin Townsley had been talking to her all this time. He now took off his helmet and tunic to make himself look less official and threatening, and managed to persuade her to let the men pitch a ladder. Talking all the time, he slowly climbed the ladder towards her. Finally at the top, they talked and talked. Then, just as she was least expecting it, he grabbed her very firmly and brought her down.

Officers of the London Fire Brigade are not given any specific training in negotiating skills, or even in the psychology of how to handle people threatening to jump. Officially this responsibility lies with the police, but in reality it is often the first fire officer in attendance who is expected to persuade the unfortunate person not to jump. In fact it is often the police who summon the fire brigade to assist with a 'person threatening to jump'.

One such call was received by the Red Watch in the early hours of the morning to a block of luxury flats in Buckingham Street, near the Strand. The police were already in attendance when the crew arrived. Apart from the policemen standing looking anxiously upward, the street was quiet and eerily empty. Halfway along the block of houses, a woman was perched precariously on the third floor windowledge. Below her was a straight drop of forty-five feet into the basement. About eight feet out from the basement was a row of iron railings topped with spikes. Martin Fittall was driving the pump that night:

The girl was young. Distressingly young, but she weighed twenty stones if she weighed an ounce. What was worse was that she was wearing a baby-doll see-through nightdress that came down to the top of her thighs, and absolutely nothing else. We took one look at her, and then at the spikes. The prospect of this great big woman hurling herself off and getting skewered on the spikes was absolutely horrible. A young copper had decided that he would use some do-it-yourself psychology to try and stop her jumping, while his colleagues attempted to gain access to the woman's flat from inside the building. He picked up a piece of tin he found lying in the street, and started to bang the railings underneath her with it. Tap, tap, tap. It was really creepy, like a big clock on a time-bomb ticking away. It seemed like ages, but must have been no more than a minute, when he stopped for a moment to talk to one of his colleagues behind him. The clock stopped . . . and she jumped. My heart leapt into my mouth as I saw her go, but no sound came out. His back was towards her as she took off: poor bloke never stood a chance. She missed the railings completely and hit him square on the shoulders and back.

On 21 June 1988 Sub-Officer Steve Short was called to a man on this tower crane, 170 feet above ground in Trafalgar Square – as high as Nelson's Column. On arrival, he tried talking to the man but was ignored, so he fixed a line and began walking along the jib towards him. When he was about twelve feet away the man stopped him, saying, 'That's far enough!' Steve asked casually, 'What are you doing up here?' The man replied, 'I've had enough of life. I'm going to jump.' Steve spent the next three hours up on the four-inch-wide jib, trying to persuade the man not to jump. The break finally came when food and drink were offered. As the man climbed down, the plastic carrier bag he was holding slipped from his grip and the contents started to spill out. 'What's in there?' Steve asked calmly. 'Everything I own,' he replied.

The young policeman screamed a terrible scream, partly through shock and partly from pain. The force of the woman's weight had severely injured his back and he lay prone on the pavement, writhing in agony. The woman, in stark contrast, stood straight up, crying, but with only a pair of bloody knees to show for her ordeal.

Sometimes jumpers jump with more tragic results. When they do the fire brigade is often called to retrieve the body from a ledge or scaffold or, more distressingly, from the spikes upon which they have often become impaled.

Trafalgar Square was the scene of another very dangerous rescue attempt by the crews of Soho's pump and pump escape appliances. It was New Year's Eve, 1982 and the usual throng of tens of thousands of revellers snaked their way across the West End, heading for Trafalgar Square. At the fire station the men of Red Watch laughed and joked and in the background the roar of the crowd cheering and singing grew to fever pitch as midnight struck. Seconds later the bells went down calling the pump to a car on fire in the Strand, near Trafalgar Square.

'It's probably a mickey,' said Sub-Officer 'Slugger' Sloane, riding in charge that night, 'we usually get one at this time.' (A 'mickey' is firemen's slang for a malicious false alarm call.)

Progress was slow through the crowds of people who thought the blue flashing lights and blasting two-tone siren were just the firemen joining in the celebration. In fact the call was genuine and the car fire was duly dealt with. Then some practical joker put in the mickey call they had been expecting, to a road traffic accident on the south side of Trafalgar Square. It was quickly established that the call was a false alarm, but the pump was by now marooned in a sea of dancing, drunken, shouting people.

Quite suddenly the mood of the crowd changed from excitement and laughter to anger and then fear. More and more people were surging into Trafalgar Square from the north side and pressure was beginning to build up on the people at the front of the crowd, who were prevented from moving forward by metal barricades. A frightened mother shouted up at the firemen, 'Help me please: take my child on the back.' The child and two women were lifted up into the back of the machine. Then someone else appeared out of the throng with his arms around another child in obvious distress. 'She's having an asthma attack, please help us.' She was given refuge in the front compartment.

Some troublemakers were now hostile, throwing bottles and full cans of beer high into the air; these came crashing down into the crowd, causing some panic. People nearest the barricades began to be squeezed tight against them. Several of the barriers collapsed and two women very near the front were pushed to the ground and crushed underfoot. A young policeman saw them go down and tried to get to them, but the crowd could not hear his shouts and he looked around desperately for assistance. Not far from him he saw Soho's pump and pushed his way through to them.

'Have you men got resuscitation equipment on board?' His voice was tense with controlled panic. At that time resuscitation gear was not standard equipment (although it is now). The crew dismounted into the chaos: Mark 'Carrots' Blakeman, John Sloane, Wally Slade and Martin Fittall.

The policeman led them to the spot where the women had fallen but the crowd were shoulder to shoulder: it didn't seem that there could possibly be anyone on the ground there, but if there was then the crowd were standing directly on top of them. Mark Blakeman was closest to the policeman:

He kept shouting at me, 'She's here, she's here', but I couldn't see anything. Suddenly I felt something soft with my boot and shoved people out of the way, enabling me to see that there was a lady lying on the ground. The rest of the crew gathered around forming a protective circle and I began mouth to mouth on the victim while the policeman administered cardiac massage. Her windpipe was blocked and I kept getting mouthfuls of vomit. We tried to clear her passage by putting her on to her front, but more and more kept coming up out of her. All this time the crowd around was trampling on us, kicking and shoving, and the rest of the crew were getting a good bashing from all sides. The young copper was so upset he started crying and begged us, 'Please don't let her die.' But it was too late.

Mark and the crew worked on her non-stop for twenty minutes, trying to breathe life back into her. Eventually a doctor arrived on the scene and declared her dead. Meanwhile Martin Fittall had pushed his way through the crowd to the pump ladder where he summoned the assistance of that appliance's crew who as a group found their way to the other woman nearby. She was already dead.

Most of the crowd were completely oblivious to the tragedy taking place so close to them and as the firemen made their way back to the fire engine, laughing girls came bouncing over to try and kiss them. Mark pushed them away: his tunic was covered in vomit and he smelt dreadful. The indifference of the crowd somehow made everything worse, the death of the poor innocent revellers more poignant and undignified.

Despite Soho's greatly overrated reputation as an area for sexual entertainment and diversion, its firemen have had their fair share of sex-related rescues, most of which are unprintable. The following two incidents are typical: the car park behind the old station was frequented by men who would have sex with prostitutes in their cars. Occasionally they parked in full view of the first floor balcony at the back of the station and the men were naturally an appreciative audience:

One summer's evening a couple were making a particularly long go of it, so when the man finally reached his peak we all gave him a great cheer and shone a large torch on to them. He jumped up out of the car and yanked up his trouser fly all in one movement. There was a stunned silence followed by a long yell – he had caught himself in his zipper like a fish on a hook. We called an ambulance for him as we didn't have any suitable equipment for such an intricate extraction.

Another time they were called to 'Man trapped upside down, Endell Street'. The front door of the house was ajar; several scruffy doorbells marked 'Model' and 'French Maid' greeted the firemen. Pushing open the door they found a man dressed from head to toe in black latex rubber, handcuffed to the first floor banisters, hanging upside down by his manacled ankles above the stairwell. Here, it transpired, he had proceeded to masturbate until all his energy was spent. He was so exhausted by his exertions he had no strength left to swing back up to the banisters. So there he was, swinging like a bat from a fruit tree, in all his glory, dizzy from the blood rushing to his head. Despite his pitiful position, the firemen were impeccably diplomatic, treating him with dignity and respect until he was out of sight, when they fell about laughing, each man adding his own unique detail of the experience.

Charity fund-raising is a very special service carried out by many stations on a regular basis, but Soho is particularly well placed for it. One year the intrepid White Watch organized a marathon car wash for the Children's Heart Ward at Guy's Hospital. With the help of Capital Radio, the event was given great publicity and on the day, the firemen prepared themselves for the challenge, armed with dozens of bubbling buckets, yards of cloth, gallons of water and unending supplies of enthusiasm.

Soon the cars began to arrive by the hundred, with a queue forming halfway along Shaftesbury Avenue. The atmosphere was filled with fun and excitement as celebrities began to line up their smart cars to help the cause. One of them was the rock star Paul Young, who jumped the queue by driving into the fire station through the automatic doors normally used by the fire engines to gain access to the street. However, these doors had been fitted with an electronic beam automatic door-closing mechanism, designed to ensure that after the machines had left the station the doors closed behind them. The beam shines across the front of the bay to prevent the doors closing on the fire engine if it is only halfway out through the door.

Unfortunately a fire engine is rather larger than Paul Young's limited edition Carrera Porsche, which sneaked halfway into the fire station, its pretty, long nose edging just under the level of the beam of light. When the unbroken beam sent its message to the great door god in the sky to close, there was no stopping it, and with a crash accompanied by shrieks from passers-by, the Porsche was crumpled and crunched by the unforgiving jaws of the dragon-like doors. The situation was defused when Paul Young's minder, an ex-Royal Marine, recognized his friend Fireman Bob Sorrell, also an ex-Marine. Tempers were quickly cooled and the great wash went on. The event raised over a thousand pounds for a very worthwhile appeal.

In the early 1970s streaking was becoming a trend. One evening Firemen Trevor Browse and Paul Grimwood were taking a refreshing after-work pint in the Avenue Bar when the conversation turned to the subject of streaks. Unprompted, one of the barmen leaned towards the two men, slapped a five-pound note down on the bar provocatively, and declared in a loud voice, 'I'll give five quid to charity if you two blokes streak the full length of Shaftesbury Avenue.' They both looked stunned. 'So will I,' another barman chipped in. 'And so will I,' a customer cheered.

The news spread like wildfire and within minutes the pub was filled with fists clenching five-pound notes. The firemen had no chance of backing out: it was a foregone conclusion before they had even spoken a word.

The following night the scene was set. It was a warm summer's Saturday evening, the theatres were full and the Avenue Bar was stuffed to capacity. Like two condemned men, Paul and Trevor set off towards Piccadilly Circus accompanied by a minder. They entered Dansey Place, which was deserted except for four Chinamen who sat around a scruffy aluminium table, their large backsides spilling over the edge of distorted chairs that looked on the verge of collapse. The portly orientals did not even glance at them until they started to strip. But by the time they had reached their underpants all four were mesmerized with fear. Then off came the underpants – the Chinamen were horror-struck and averted their eyes back to their now less appetising meal.

The boys were under starter's orders. 'Wait for it,' their accomplice growled like a parachute regiment sergeant major. 'Go!' he roared, and they were off.

The timing was impeccable. All along Shaftesbury Avenue the theatres were disgorging thousands of people into the street. In the Avenue Bar, the air was hot with expectation and other firemen were carrying buckets around, collecting more for the daring duo. The customers on the pavement whistled and cheered as the two athletes approached the finishing line, Fireman Basil Brazier's candid camera flashed and women gasped. The naked bodies flew headlong into the bar and into the arms of the law. Two large police officers looked down at the men disdainfully. ''Ello, 'ello 'ello, what's going on'ere then?' asked one theatrically. But all was not lost: they had been briefed and decided to add to the fun with a fire-service-style 'wind-up'. Thanks to the streak more than £150 was raised for charity – 'and that was quite a lotta dosh in 1974!'

Cars leaking petrol, gas leaks, explosions in electrical junction boxes, chemical spillages and road traffic accidents are all special services that the men of Soho have to deal with, although calls to road accidents are not common on Soho's ground – there are few roads where people are able to get up enough speed to have really dire consequences if they crash. Nevertheless Soho has its fair share of accidents: a person can be trapped in a car even if they are only travelling at thirty or forty miles an hour when they crash. And there are a few parts of Soho's ground where crazy people drive at two or even three times the speed limit, including the Embankment, the Mall and Pall Mall.

Leading Fireman Vernon Trefry surveys the grisly aftermath of a road accident where a car has crashed into the pedestrian walkway leading from Hyde Park Corner underground station.

Three of Soho's five adjacent stations have many more accidents on their ground – Knightsbridge Fire Station with the Piccadilly underpass and Park Lane, Euston and Manchester Square Fire Stations, with the fast-moving Euston and Marylebone Roads – so it is not uncommon for Soho's machines to be called on to neighbouring grounds to deal with serious car accidents. However, the overall number of traffic accidents that central London fire stations attend is minimal compared to their suburban rivals with major motorways on their grounds.

Fire engines have road accidents, too. More than once Soho's turn-table ladders have ended up in the front of shops or in the basement rooms of private houses, but considering the number of calls to which they are required to respond the number of accidents (or 'biffs' as the firemen call them) is mercifully small. However one accident worthy of note was perpetrated by 'Mad Dog' Mowlem who swung his appliance across Shaftesbury Avenue and put the top end of the turn-table ladders through the back section of a double-decker bus, trapping all the passengers on the top deck; they had to be rescued using a ladder off the pump.

In keeping with their Thames watermen heritage, the men of Soho Fire Station are responsible for the northern half of the river Thames which flows along their southern boundary. Soho's crews are not often called to emergencies on the river but the West End was once again the focus of world press attention when a pleasure cruiser called the *Marchioness* was sunk by a barge late in the evening of 20 August 1989.

Red Watch with Station Officer Bruce Hoad were on that night and were the first appliances to arrive on the scene. A major incident pre-set procedure for river accidents was launched by Command and Mobilising Centre as soon as the first call was received and by the time Soho's machines pulled up on Victoria Embankment police launches were already landing survivors on a pier. Bruce Hoad ordered one of his appliances on to the next bridge down river of the incident, to set up spotlights on the water to look out for people who might have been swept along in the current. He requisitioned a boat and set off to assist in the search for survivors in the water.

They searched for hours without finding a single person. All around them the water was filled with bobbing tragic souvenirs of the party which had been taking place on the boat. Fancy chairs floated by in the murky darkness; an evening shoe, a jacket, a life-belt . . . but no people, alive or dead. Those who had survived all made it to the shore in the first few minutes after the ship sank.

Fire appliances from all over London were mobilized to assist with the search and rescue operation, and men set up search parties with spotlights on every bridge between the West End and Putney. Overhead, the roar of a Royal Air Force Sea King helicopter added a greater sense of drama to the surreal scene taking place in the very heart of London. The firemen excelled themselves that night, but to little avail. Once what remained of the boat was lifted to the surface they were faced with the horrid operation of retrieving bodies from the wreckage. The only recognition they received for this dangerous and upsetting task was the knowledge of a job well done.

Perhaps it is with animals that the most outrageous and unpredictable things happen. The adventures of Colin the Crab encapsulate the spirit of Soho Fire Station today: Colin was a splendid scarlet crab rescued by the Blue Watch at an incident at a Chinese restaurant in Gerrard Street in 1986. No-one had the heart to cook Colin, so he made his home in the kitchen sink – which was fine, except that he died within a couple of days. The whole watch were crestfallen and reluctant to dispose of the corpse, but the smell soon forced the issue. A few days later, a large plastic substitute crab with bulging eyes the size of ping-pong balls was presented to the watch. Colin became an instant celebrity and soon began to take trips abroad with various members of the station. Within a couple of years he had visited most of Europe and had even been all the way to Australia. He never once failed to send a postcard to 'the boys' and he features in dozens of holiday snaps: lying by the pool, drinking a gin at a beach-side bar, surfing, disco-dancing: whatever was happening, Colin would be involved. When he was not on holiday he would sit grandly on top of the television in the mess room, staring down on all he surveyed with his bulbous eyes. He was by now part of the family and held in high esteem.

Then a vicious and scurrilous plot was hatched by another watch. One morning the Blue Watch came on duty to find a ransom note for Colin: he had been kidnapped and the demands were high. The stakes were higher still, for if the kidnappers' demands were not met one of his claws would be cut off and sent through the post.

News of the dire situation spread throughout the whole division but alas, poor Colin, the Blue Watch did not take the hardened criminals of the Green Watch seriously and within two days a large but disembodied plastic crab claw arrived on the office desk. Suddenly things were very serious indeed: the special branch were mobilized and after a daring dawn SAS-style raid, Colin was found blindfolded in Fireman Dan Shilabeer's locker and returned to his rightful place in the mess room. 'Of course, Dan had had to be tortured good and proper before he would hand him back,' said one of the victorious Blues. Since then Colin has led a quieter life, declining foreign holidays and living for the most part in the false ceiling above the dining table in the mess.

One dark and rainy night the bells went down, summoning Soho's turn-table ladder to Portman Square to assist fire crews from Manchester Square to retrieve a parrot from a tree. The drama had started some time earlier in the evening when the owner had taken the parrot on its regular evening stroll. Not for the first time, the parrot, named Dracula, decided to taste freedom and took off into a nearby tree. He usually returned to his owner when called, but had already required the fire brigade to persuade him back down on a previous occasion. This particular night he was proving very obstinate. The police had been summoned and in turn had called the RSPCA. A nice man had eventually arrived armed with a large perch on the end of a long stick. He waved this up and down temptingly in the rain under Dracula's tree. Dracula ignored him. Then the owner stepped forward and added usefully, 'By the way, he's blind.'

The by now exhausted RSPCA officer was led away, whilst the policeman in charge, head in hands, ordered the fire brigade to be summoned. The appliances from Manchester Square were rapidly on the scene but their turn-table ladder malfunctioned and so Soho's were ordered on by Wembley Fire Control (the mobilizing control was based at Wembley before CMC opened in 1989). Eventually the ladders were pitched, stopping just short of Dracula's perch. Leading Fireman Micky Thomas climbed up towards him, reached the top, and then, in Wally Slade's famous 'Donald Duck' voice, held his hand out, muttering, 'Come on, Dracula, come to Daddy!'

On the ground everyone held their breath as Micky inched forward towards his prize. Two feet, one foot, six inches. He lurched out, trying to grab him, but to no avail. Dracula took off in all his splendour, like one of his namesake's bats. Unfortunately, without the appropriate radar system, he was risking a lot and within seconds he collided with a great thwack into the side of a nearby building. With a squawk and a flutter of feathers he plunged to the pavement, where he landed like a 'bag of bricks'. Before he could move three firemen rugby-tackled him simultaneously. Against all the odds Dracula survived intact, and was returned to his delighted owner.

Firemen Tom Smith and Steve Davies were riding the turn-table ladders when called to 'Cat in precarious position, Goodge Street', where Station Officer Staynings was already in attendance with the pumps crew. The cat had climbed on to a parapet at the sixth floor level of the building and its distraught owner stood desperate, wringing her white-knuckled hands and biting her lower lip nervously. It was quite impossible to reach the cat from the nearest window.

The ladder was pitched and Steve Davies began the long climb. Tom Smith, who at ground level could not even see the cat far above, turned to the distressed lady and in his most serious but calming voice said, 'Madam, I can see that your cat has been under a great deal of stress recently, but there is no cause for alarm: that fireman is a fully trained professional cat psychologist. If anyone can talk your cat down, he can.'

By now a large crowd had gathered, all talking excitedly amongst themselves. On reaching the parapet Steve found the poor cat cornered, eyes wild, back arched and claws drawn in terror. Near to it lay the broken head of an old broom. He reached forward and grabbed it. The crowd shouted, 'He's got it, he's rescued the cat. Hooray!' The next moment Steve lost his hold on the brush: it slipped from his hands and nose-dived towards the pavement, the owner screaming long and loud.

'Don't panic, Madam, it's a decoy!' Tom reassured her.

The Guvnor walked away, his head in his hands: he had only just come to Soho, and already they were living up to their reputation as champion pranksters. The real cat was quickly carried to the ground by Steve, and returned to its grateful owner none the worse for the adventure.

The best story of all about an animal rescue took place off Soho's ground. Station Officer Dan Ivall was in charge and 'Turk' Manning was a junior hand in his crew:

I was still a young buck with Frank Bartley – we got into all sorts of scrapes, including a time when Dan Ivall had to get us out of prison after we were arrested for climbing over the back wall into a nurses' home, having mistaken it for the back wall of Soho Fire Station.

One cold winter evening we went up to this fire on Kentish Town's ground. The fire was on the ground floor, and the smoke had poured into the first floor flat. The distressed owner implored, 'My budgie is up in that flat, please rescue him.' We dashed into the smoke on our mission of mercy. The budgie's cage was on a high stand, suspending it in a thick cloud of deadly smoke – all seemed lost.

They got his limp body out, his yellow wings all sooty and greasy, and laid him on a table below the level of the smoke. Fireman Terry O'Sullivan, a big man with fingers and thumbs like a bunch of bananas, began artificial respiration, pumping away at the little wings just as if he were trying to get the water out of a drowned man's lungs. But it wasn't working: drastic action was needed. There was only one thing for it: oxygen.

While one of the crew dashed out to fetch a spare oxygen cylinder (part of the standard proto breathing apparatus), the others stood around the 'death of Nelson'-like scene, the little blackened bird laid out on the table, the room cut in half by the horizontal wall of smoke, and in the fireplace, a warm, welcoming log fire, crackling away merrily, oblivious of the damage its inhospitable cousin had wreaked in the flat below.

In the early seventies the resuscitation of victims overcome by smoke was very primitive compared to the sophisticated techniques now available to fire and ambulance crews. Turk Manning continues the story:

If someone had collapsed from smoke inhalation but could not be given mouth-to-mouth resuscitation, we used to put a fire helmet down next to the victim's face and 'crack' an oxygen cylinder into it, so that it provided a concentrated supply of oxygen filling the volume of the helmet. I handed the cylinder to Frank, who was almost overcome with emotion. He struggled for a moment with the cylinder but in his excitement, instead of just cracking it, he opened it fully, sending a great jet of oxygen towards the bird. In one great whoosh, the budgie, which had just been showing signs of life, took its last flight across the full length of the table top, straight into the burning fireplace and on to meet its maker. Which just goes to prove the old fireman's adage: 'If it's got your name on it, you're going to get it!'

One of the golden rules at Soho Fire Station has always been 'Work hard and play hard.' By its very nature the work of a fireman is stressful – no-one knows when the bells will go down, and when they do, what the men will be called upon to do. Firemen do not have time to laze around, either. Contrary to the condescending public image of them as a group of overgrown schoolboys playing cards and snooker all day, the reality of life on a modern London fire station is one of almost continuous work both on and off the station in between responding to emergency fire calls.

The rota system they operate divides the complement of men into four watches, Red, Blue, Green and White, working a four-day shift of two day duties from 0900 to 1800, then two night duties from 1800 to 0900 and then four days off. On the day duties the men are required to carry out numerous station routines, including health and safety checks, cleaning of machines and testing of equipment. They must drill every day, keep individual log books of their drill and training record, and do regular physical fitness and strength training in the multi-gym at the station. In addition they have to carry out numerous visits to buildings on the ground for fire safety, prevention and certification checks and to familiarize themselves with buildings of particular note or complexity. (These familiarization visits are known as 11D visits after the particular paragraph of the Fire Services Act; Soho has hundreds of buildings requiring such visits within its one square mile patch.)

The firemen must also carry out checks on every single water main hydrant – of which there are thousands. Each hydrant has to be visited by a crew armed with a metal bar similar to a crowbar and a standpipe which is screwed into the supply pipe under the pavement cover to release the water. Many of the hydrants become damaged or fall into disrepair and written records of the state of each hydrant have to be made and forwarded to the appropriate authority.

Other buildings have to be visited to record particular 'risks' – hazardous chemicals stored in properties, buildings under development with only partial floors constructed, biological hazards in hospital research wings including deadly stores of contagious and infectious diseases. And all of this at a station which responds to an average of one emergency call an hour, twenty-four hours per day, 365 days of the year. Each call brings a plethora of administrative work to the officers of the station which must be completed as soon as the appliances return. As it is the station office has a nightmarish continual flood of forms and documents to be filled in, copied, filed, signed . . .

It is therefore not surprising that in the few periods that the men do have to relax during the day or late in the evening they get up to mischief. Spontaneous bouts of boisterous play relieve tension and forge bonds between the fire-fighters which are vital for a 'happy ship'. For an employer to think that a group of men (and now women, too) can be expected to spend so much time together and not get up to tricks and pranks is foolish, yet senior officers are clamping down on the fun and games, killing off the final link with the old spirit of the job. Firemen love practical jokes and are particularly creative and daring in carrying them out. It is something to do with the very nature of the job: some of the most notorious pranks have been perpetrated by Soho's firemen and it will be a shame if they must now pass into the ever-increasing list of 'past glories'.

The Indian rope trick was always very popular with the crowds: one of the firemen wrapped his head in towelling to form a turban, draped an old salvage sheet over his shoulders and then sat cross-legged in front of a bucket containing a line bag (a canvas sack storing a long rope). In his hands he clasped a train warning horn (carried on the appliances for use on railway lines to warn trains of obstructions). A fishing line was attached to the end of the rope and passed to an accomplice on the balcony above. Once the rest of the firemen had hidden in various vantage points out of sight of the main road, the show began. The mysterious Sinbad-like figure charmed the snake to appear out of his basket and dance a wonderful wistful jig, accompanied by the haunting wailing of the 'eastern' pipe. Once a really large crowd had gathered, the charmer would stand up very solemnly, take a long, low bow and disappear into the darkness of the appliance bay behind, the doors closing like the curtain of a theatre, leaving them chanting and cheering for more.

Other favourite pranks included placing a tempting-looking football filled with sand on the courtyard in front of the station on Cup Final night when the West End filled with drunken supporters, every one of whom fancied himself as the best centre forward in the world. One by one they would line up to kick the ball all the way to Piccadilly Circus, only to find this particular ball had teeth and bit back! Putting concentrated foam mixture down the loo was always good for a crack as well, especially with a new recruit. One flush was normally enough to produce about fifty cubic feet of thick white bubbles!

Any time a new face arrived at Soho Tony Cooper did his old wino trick. He'd dress up in a scruffy cap pulled down low over his face and a dirty old black raincoat, and carry a half-full wine bottle and a walking stick. Like all the best pranks, it always worked and was just as funny each time, not least because of the variety of ways people reacted to it. Station Officer Bob Fielder was a very powerful man who was not afraid to use his considerable strength: when he was confronted with Tony's tramp, he simply picked him up by the throat and carried him out of the station. Lenny Goodfellow was subjected to a variation on the theme when Tony actually got into his bed. 'I'm not sleeping in that bed, Guvnor!' he whined. 'Oh, yes you are, sonny' the Guvnor barked back at him.

When John Peen came to Soho as Guvnor, the men decided that he too was worthy of their favourite prank. Knocker went into the station office and told JP, 'We've found this tramp in the basement, Guvnor, what shall I do with him?'

'I'll come down and have a look,' he replied, reaching for his undress cap and placing it smartly on his head. They showed him into the basement and there in the corner, gurgling and swearing under his breath sat this nasty old tramp. JP gently kicked him a couple of times and gave him a crisp order. 'Go on, get out of it.' Then he turned to Knocker and said, 'Cor, don't he stink,' at which point they all burst out laughing. To distract JP from the laughter Tony whacked him across the knees with his walking stick. That, of course, made them laugh all the more, and the game was soon up.

This particular trick had its greatest triumph when Tony fooled the entire watch. They were on the balcony doing another of their favourite tricks: lowering a large rubber spider on a string on to unsuspecting passers-by. Tony sneaked away unnoticed, dressed in his costume, slipped out of the side door and then hobbled along the pavement towards the station. The effect was magnetic, like a large juicy salmon swimming into the view of ten fishermen, the irresistible bait seemingly exactly in its path. Stifling giggles of anticipation, they waited for the right moment . . . 'Bombs away,' whispered one as the spider was dropped. It landed just inches in front of Tony, who instantly grabbed his chest, let out a loud gasp and collapsed to the ground. The effect was electric: never in the history of the known world have so many firemen disappeared so quickly.

'I nearly had a heart attack,' Knocker White said afterwards, 'especially when the two coppers came across the road and rang the station bell to tell us they'd seen a very angry looking tramp keel over and then run into the station yard brandishing a walking stick. We helped them search the station, but they never did find the tramp.'

One day several members of the Blue Watch telephoned one of the public call boxes outside the station 'just as a fancy dolly bird was coming along the pavement'. She answered it, so one of the firemen said in a thick Texan accent, 'Hello, love, I was wondering if you would like to be in my latest film?'

'Oh yes!' she replied excitedly.

'Well listen, honey,' he smoothed, 'I'm in Wingate House on the opposite side of the street, would you show me your best raunchy walk as an initial audition?'

Without further ado she set off along the road, swinging her hips like Mae West, then ran back to the phone and said, 'Is that all right?' There was no reply because the men were all doubled up with laughter. Foolishly they could not resist having another peek at her over the window ledge. Suddenly she heard stifled giggling and snorting and looked up to see all the men at the window like naughty schoolchildren.

One of the most bizarre of recent special services occurred in a mortuary in Euston. The Red Watch were on duty. The moment Fireman Bob Moulton read the call slip, as it printed out in the watchroom, he knew this was going to be a 'bad one'. Euston's pair (the pump and pump escape from Euston Fire Station) were out on an exercise, so Soho's pump was standing by and went instead, along with the emergency rescue tender, also from Euston. The circumstances leading up to the call had been a series of disasters from the outset. A sad, solitary middle-aged woman with no relations and few friends had died in her tenement block flat. She was not found for several days and as a result she started to decompose and swell up. The council had sent around a couple of men from the undertakers but they weren't able to lift her, which was not surprising as she weighed twenty-eight stones! When they returned the next day they realized they could not get her into a coffin. She had swollen up to twice her already vast size, so the undertakers had to perform a type of surgery to release the natural gases which had built up inside her due to the decomposition of her body and manhandle her into a large bag.

Fireman Bob Moulton related the story to one of the younger firemen at Soho who hung on every word as the story progressed:

Once they got her back to the mortuary, quick as possible they jammed her into the fridge. Well, that was it, weren't it. She just sort of, you know, slumped open: spread out over the weekend! On the Monday morning the mortician came into work. He was a skinny, pale-blue-skinned man with long, thin, white fingers and deep-set eyes like a skeleton. He opened the fridge to have a look at his new arrival, but when he tried to pull her out he found she was jammed fast. And what's worse she'd gone bad: turned to rancid meat, all purple and green.

After a short deliberation the mortician decided to telephone the fire brigade for help. He greeted the crew at the door with a crooked-toothed smile, rubbing his hands together and saying, 'You're going to love this one, boys,' just like someone from a Hammer House of Horror movie. First of all the special gear was unloaded from the emergency rescue tender to try and dismantle the fridge around her, but that didn't work. Every time anyone touched her body, this foul-smelling gas came puffing out. Colin Birch from Euston, known as 'Lurch' thanks to his particular style of driving, climbed right in on top of her to try and drag her out, but that didn't work either. In the end they managed to get a salvage sheet around her with a line:

Then we grabbed the line tight, literally pulling her back together. As we pulled her the gases came squeezing out. The smell was terrible. Every time we moved her a bit, she hissed and we were terrified of dropping her. As she slowly emerged from the fridge she almost fell on to the floor and I found myself grabbing on to her flesh to keep her up. Eventually we got her on to a stretcher and that was that. God, I was glad when that was over.

Once the men had recovered from their exertions, they had a look around: in another room there were four or five bodies stretched out on the tables just like a car body workshop – all the bodies with nothing inside them. Then the mortician said he had something special to show them: 'Arrived in the post this morning. Here, have a look at this,' he said, with a smile, offering them what looked like a large cake box. It was the head of an unidentified victim killed in a fire. Said Bob afterwards, 'The mortician was a funny fellow; had no flesh on him. Still, I suppose if you worked in a place like that you'd probably never eat, would you?'

God willing Soho will never again see explosions like those she suffered during the Blitz, but the endless warring factions of rival criminal elements in the area meant that during the fifties and sixties explosions caused by various devices were not uncommon, and as always, Soho's firemen would be in the very thick of the aftermath of such urban battles.

In the early 1970s the Irish Republican Army moved its bombing and terrorist campaign to London and numerous explosions took place in Soho and the West End. At many of the incidents, like the sad explosion at the Wimpy Bar in Oxford Street where a police bomb-disposal officer was killed, the brigade were not actually required to fire-fight, but were on stand by in case a fire broke out and to assist in any other way they could.

One night the IRA decided to change its tactics and set off a dozen incendiary devices in various buildings along Oxford Street. Fireman Tony Cooper was with Sub-Officer Steve Fletcher on the sixth floor of one of the office blocks involved. The fire had burned away the staircase around them. Sub-Officer Fletcher turned to Tony and said, 'I think I'd better go and get a ladder,' but the words were hardly past his lips when the floor he was standing on gave way and collapsed. He disappeared from sight with a puff of smoke like Aladdin's genie vanishing through a pantomime-stage trap-door. Luckily the landing of the fifth floor held his weight, but afterwards he said, 'It was a nasty moment because I thought I was on my way down to the ground floor!'

That night became known as 'The Night of the Bombs' as the crews worked all night: Soho had never had so many serious fires burning simultaneously since 1941. In the last twenty years, explosions have occurred in numerous other locations including the YMCA in Tottenham Court Road; the Oasis swimming pool in Shaftesbury Avenue; a post box in Piccadilly Circus; Lower Regent Street; Old Compton Street; and the Army and Navy Club in St James's Square.

Perhaps the most ghastly bomb call Soho ever picked up was to a 'human bomb'. This particularly unpleasant incident occurred at the men's hostel in Endell Street. One of the inmates left a note in his room saying that he was fed up with life and was going to end it all. He had gone into the loft where the lift motor room was, through a small hatch in the ceiling, fitted with a trap door. He sat down on the floor and set the room alight. The brigade arrived and tried to gain access to the loft, but as they opened the door, he shouted, 'Clear off, you bastards, I've got a bomb! I've got a bomb tied around my waist. Clear off!'

The room was already quite well alight. The police were summoned and a jet set in. When the jet arrived a fireman tried to re-open the hatch:

I glanced into the loft and could see the ghastly spectre of this bloke

White Watch firemen climb through the blasted out windows of the Army and Navy Club, St James's Square, seconds after arriving at the scene of an IRA terrorist bombing.

running around alight. His clothes and hair were burning. He stumbled towards me, tripped and then fell with a crash on top of the hatch door, which slammed shut. As a result we could not re-open the door. And there he lay, burning away just the other side of where we were.

Eventually they cut away the frame around the hatch and with a great crunch the remains of the man came through the hole. Once he was out of the way the fire-fighting was quite straightforward and the flames were quickly extinguished. It turned out that he did not have a home-made bomb attached to him: he had just used that as a crazy excuse to keep them away from him.

Explosions are not always the result of terrorist bombs. Two careless builders were transporting a tar boiler in the back of a flat-bed lorry used by roofers. They also had propane cylinders in the back of the van. Recklessly they kept the tar burner alight in the van driving between jobs. They had just come past the Oasis at the top of Endell Street when it caught fire. The builders noticed it and jumped out, screaming to nearby drivers to run for their lives. A minute later the cylinders exploded with a great thunderous roar. The van blew into a thousand pieces and many of the nearby cars lost their windows, while inexplicably other cars right next to the explosion were undamaged. The other cylinders had to be cooled and immersed in a dam. As the incident progressed the full extent of the damage caused by the blast emerged. Buildings up to three hundred yards away had had their front doors blown clean off their hinges, yet others very close to the explosion were untouched.

1990 saw Soho's ground as the centre for another terrorist bomb attack. The Carlton Club in St James's Street was bombed at lunchtime on 25 June. Station Officer Chilton, White Watch, arrived first on the scene and organized his men in the most daring and hair-raising rescue operation. Most of the ground and first floors had been blown to smithereens. People were trapped in the burning building. Ladders were pitched across gaping holes in what remained of the floors. In all twenty-three people were rescued, while the injured were ferried out into the street and on to waiting ambulances. The rescue operations were still in progress when reports emerged of a possible second explosive device in the street outside, but the firemen stuck to their principal task of rescue despite the great danger they faced. Once the building was cleared of survivors and the fire was extinguished, all the crews were withdrawn to a safe distance while the police bomb-disposal teams dealt with the additional suspect device.

In 1991 the IRA launched a mortar rocket attack on Downing Street (which is on Westminster's ground) but the van used in the attack was parked in Whitehall Place, which is Soho's ground. The van was severely damaged in the attack and burst into a ball of flame. However the danger of further explosions was such that the firemen were not actually allowed close enough to the burning vehicle to extinguish the fire and it eventually burned itself out.

Soho has witnessed many bizarre incidents. Through their years of service firemen are subjected to horrors of human tragedy that surpass most ordinary men's imaginations. Whatever thoughts of glamour and excitement go through people's minds as they witness the splendour of a fire engine racing through London's streets, lights flashing and siren blasting its warning, one should remember that they are usually on their way to help someone in great distress.

Firemen's tales are often grisly and cruel. The realities of suffering and death are sometimes more than a listener can stand. But no amount of handling 'stiffs' fried in a fire could have prepared the men of the Green Watch at Soho Fire Station for the horrors of the night of Friday 15 August 1980, 'the night of the 37 Club'.

Dante's Inferno 7

In and around Soho, there are still a multitude of derelict, unoccupied, run-down premises used as strip joints, gambling dens, illegal drinking bars and clubs, operating without licence and often including sex and drugs among their attractions. Their existence is usually known to the police but on rare occasions they are allowed to remain open for as long as they provide the CID with a useful mine of information on the still thriving underworld: the persons who frequent these dives are in the main crooks, pimps and drug dealers. As a result, the police do not always inform the local authorities or the fire brigade of the location of such premises. Under just such circumstances two club premises operated in Denmark Place, a typical little dead-end street off Charing Cross Road. The police had raided the clubs on several occasions and succeeded in closing them down, but they soon reopened under new management.

In mid-August 1980 the two clubs had been placed under surveillance and on the night of Thursday 14th, two or three undercover detectives, including a woman police officer, were on the premises: they intended to raid and close down the clubs the following Monday. Number 18 Denmark Place was a three-storey terraced Victorian building, squeezed in amongst a hotchpotch of buildings and situated about halfway down the cul-de-sac. It backed on to shops and dwellings in Denmark Street. The building was of traditional construction: Victorian brick walls lined with plaster, the ground floor made of concrete, and all the other floors timber. The ground floor was open plan and rented out as a garage.

The clubs were on the first and second floors of the building. They were known as the South American Club or 'Rodo's', and the Spanish Night Club or 'El Hueco', 'The Hole'. Entrance to the clubs was via the front door in Denmark Place. This door led into a small entrance area and then to a flight of wooden stairs. The stairs were enclosed and led up to the landing area of a rear wrought-iron fire-escape staircase. Off this landing

View of the club building showing doors, stairs and exits.

was the entrance to the first floor club. The fire escape then continued up to the second floor, providing access to the upper club. The portion of the fire escape leading from the first to the ground floor was open to the elements, but the upper portion was enclosed with plywood, providing shelter from the rain for customers moving between the two clubs.

The existence of the clubs was not known to the London Fire Brigade or the local authority. The last time a Fire Prevention Officer had visited the premises was in 1978, when a music shop in Denmark Street had taken over number 18 Denmark Place, at that time almost derelict, to use as a hostel for some of the young hopeful musicians who so often frequent Soho. As a result the Fire Brigade had insisted that a fire escape be constructed. After a short time the music shop in Denmark Street had given up running the hostel; the place had descended further into disrepair and was no longer subject to fire brigade inspection.

The fire escape was badly rusted with steps missing. At its base rubbish collected in rotting piles, the sulphurous stench hanging in the air. The supposed escape route door into Denmark Street at the bottom of the stairs had been bolted and barred. The upstairs windows facing on to Denmark Place were shuttered with wood to stop people looking into the clubs and to prevent them being broken by rowdy customers. The front door, which was the only entrance, was steel-lined and secured with a heavy lock.

One of the many dubious characters who frequented this club was a petty East End crook named Joe Thompson, known to his associates as 'Gypsy'. He was a loner and a dreamer who thought that everyone liked him and used to think of himself as one of the boys in the big league.

He lived on the upper floor of a house in the East End, letting out the downstairs rooms. In the spring of 1980 he had rented out one of his rooms to a young couple. During the course of one evening he heard them having sex. Without hesitation he went downstairs, knocked loudly on their door and asked to join in. When they refused, he returned to his room, drilled a spy hole in the floor and briefly found satisfaction watching the couple from above. But minutes later, his blood roused and frustration heightened, he went down again, apparently using the threat of eviction to persuade them to let him in. The details of what happened are not clear, but apparently he did join the other two in bed and was thrown out for 'overstepping the mark'. Following the subsequent argument he went back to peeping. The couple, seemingly unperturbed by these events, continued their love-making.

In the end Gypsy could stand it no longer. He seized a can of lighter fuel, went charging down to the landing below and poured it through the letter box of the couple's flat. He waited a few seconds, thrust in a lighted match and returned to his own room. The smell of smoke soon filled the house. The man opened the bedroom door, saw flames, panicked and jumped straight out the first floor window, landing in the street with a terrible crunch, breaking his leg. The girl escaped via the hallway, suffering slight burns as she ran through the flames. The fire brigade were called and the fire was quickly extinguished. All this time Gypsy was in his room, his eye pressed to the floorboards, leering through the hole he had cut.

The unfortunate young man, frightened that Gypsy would evict them, told the police that the fire was his own fault, that he had accidentally poured lighter fuel on to the floor while filling his lighter. Despite this explanation, the fire was recorded as doubtful by the fire brigade officer at the scene . . .

During the evening of 15 August, both clubs in Denmark Place began to fill up with clients. It was a very warm Soho night, the muggy air still and sticky, heavy with the smell of rancid rubbish, clogging traffic exhaust, frying onions, the rotting corpse of a rat bulging in the gutter outside. Inside the first floor club it was steaming hot, the condensation and sweat of its members dripping down the walls. The undercover police officers had been withdrawn from the premises the previous night. The doors of both clubs that led out on to the fire escape were open, especially important as the windows were boarded up. The people inside were of all types: 'pimps, prostitutes, drag queens, drug dealers and other undesirables'. The majority were foreign, mainly South Americans, and many were lonely illegal immigrants who went to the clubs seeking company.

Membership of the clubs was strictly by introduction: once accepted, members were issued with a card stating their name and a number. On a good night up to 150 members would squash into the sleazy den. This was the case in the early hours of 16 August 1980.

Neither club employed a doorman or anyone to check who came in and out of the clubs; means of entry was like something from a 1920s prohibition gangster movie. Upon hearing the doorbell, one of the staff would look out the upstairs window and ask the person seeking entry his name and membership number. When they gave the correct information, or were recognized as a regular customer, the front door key would be thrown down.

Just after midnight Gypsy arrived. The usual first floor window opened and the shape of a man appeared, his features hidden in the dark shadows. The face recognized Gypsy, fumbled for a moment, then dropped the key down to him. Gypsy let himself in to the darkened, musty corridor, turned around in the gloom and, as usual, locked the door behind him.

On entering the club he approached the bar and ordered a rum and

Coke. He was already drunk. He handed over a five pound note in payment and was given no change. He started to argue with the barman who told him that he was not getting any change. The barman warned him, 'If you cause trouble, you'll be out.' Somebody else came over to him, saying, 'Your girlfriend was here earlier, but she's gone home with some other bloke.' This information upset him still more. He was also told that another club member was after him because he had cheated him on a deal.

He gulped his drink, went up to the second floor and pushed his way through to the bar. Almost immediately he was warned that a man was looking for him. 'You sold him some fake jewellery and he wants his money back.' Gypsy exploded, his gruff, frog-like voice rising in pitch. At that moment a tall, rough-looking character approached him and demanded his money back for the fake jewellery. Onlookers intervened, preventing an all-out fight. Gypsy was escorted down the stairs and thrown out into the street. This was at approximately 0230.

Gypsy left the club in a rage, screaming abuse and threats and punching the air with his fist. In drunken fury he decided to get his own back by using his favourite trick.

Hurriedly he searched the area outside the club for a suitable receptacle. He found a two-gallon plastic container. He walked to Charing Cross Road and hailed a taxi, telling the driver that he needed to find a petrol station to refuel his car. The cab took him all the way to a station in north-west London. Though it was illegal at the time, Gypsy managed to obtain petrol. The taxi driver should also have known better: the law prohibited him carrying a passenger with petrol in a plastic container, but he delivered Gypsy back to Denmark Place.

It was now almost an hour since Gypsy had been thrown out of the club. Remarkably, his temper and the effects of drink had not cooled enough to make him change his mind. He reached the entrance of the club at about 0320. The street was empty and silent save for the cocktail party gurgle of voices, the subdued heavy beat of the music and the sound of laughter filtering out from inside the club.

'It's full,' he must have thought to himself, 'Now I'll give them something to think about.' Grasping the container, he began to pour petrol through the letter box in the door of the club, but the angle of the neck caused more petrol to be spilled outside the door than inside. Calmly he searched for something to use as a spout. Another piece of plastic served the purpose. He poured the contents of the container into the hallway. He reached into his pocket and tried to ignite the fuel with his cigarette lighter but failed repeatedly.

Whilst he was trying to light the fuel a man entered Denmark Place. He passed on his way. Gypsy waited until he was out of sight, then returned once more to his task. About five minutes had passed. The petrol was beginning to vaporize: a slight draught from the ground floor carried the vapours up the stairs and in through the open doors of the two clubs above. The petrol smell grew stronger, but the cigarette smoke in the rooms above, combined with the smell of sweat and scent, was more powerful still. Frustrated by his failure with the lighter, Gypsy lit a scrap of paper, pushed it through the letter box and ran off.

Above the music, the clink of glass on glass, the laughing, coughing and sick sighs of false delight, a rushing sound, the sound of a great wind, like the roar of a dragon, imposed itself. For a split second time stood still, then with a great fire-spitting gasp, the petrol vapours ignited like several large blow-torches. Fuelled by nearly two gallons of petrol in a very confined space, the flames shot up the stairs, igniting everything in their path as they burst into the crowded rooms. Feeding on the vapours that had already spread throughout the premises, flames engulfed the entire building like a furnace within a couple of seconds, catching the occupants completely by surprise.

As the fireball exploded up the stairs, panic spread within the club. It was as if the Four Horsemen of the Apocalypse had arrived: those able to move fast enough ripped the shutters from the barred windows, tearing the skin from their hands, then breaking the glass with bleeding knuckles. Some of them trod on those who had already fallen. Some jumped to the ground, landing with a crack as bones broke in the fall. Some threw themselves from the windows, their clothes ablaze. On the stairs chaos prevailed as people tried to descend against the rising fire. Others stumbled blindly head-first into the inferno. The lucky ones were standing on the lower section of the fire escape where they had sought refuge in the deceptively cooler night air. The explosion of hot gases physically pushed them down the stairs.

Len Hazell was sitting near the door. His mind raced: 'Everyone is running . . . screaming . . . oh my God . . . oh my God . . . their clothes are on fire . . . their skin is peeling off their faces . . . the screaming . . . the screaming . . . the screaming.' He found himself outside, not knowing how he had made it. There followed a terrible silence.

Other people made it to the rear courtyard and tried to escape via the back door but found it locked. The ground was littered with old tyres and other rubbish, all of which began to smoulder as the sparks rained down. Those who made it through the thick smoke to the bottom of the stairs were then showered with burning embers and debris from the fire above. On finding their escape blocked, many climbed a high wall and got away via the adjoining premises. Others broke the rear windows of the shop, cutting themselves terribly. Yet even then their trial was not over: at the front of the shop they found the windows were secured with grilles. They tried unsuccessfully to smash their way out through the windows.

By now the fire was punching out of the windows and into the well of the building. The plastic drainpipes were melting in the reflected heat, contorting, twisting and dripping to the ground like blackened treacle. Inside the building victims met a similar end.

The firemen of Green Watch had been on since 1800, and it had been a particularly busy spell of duty.[1] Apart from the normal routines, they had been called out on numerous incidents. They returned from a previous shout to the station at about 0245. By the time they had re-stowed their appliances and completed the necessary office paperwork, it was not until about 0315 that they were able to relax. As so often, the talk turned to fires. Turk Manning, the Station Officer in charge, and one of the most professional and experienced firemen in the brigade was reminiscing:

'In the good old days we used to get so many club fires – most nights we could almost guarantee a good working fire at the height of the gangland wars – it's been years since we had a club fire.'

The words were still cool on Turk's lips when crash, down went the bells, shattering the quiet of the station, awaking in everyone a spine-tingling sense of urgency, anticipation and excitement. As they made their way down to the appliance room, the clatter of the teleprinter chattering out the address of the incident could be heard clearly. The time of this call was 0333. The duty man pulled on his leggings and peered at the printer roll, to read:

Fire, Denmark Court, Off Charing Cross Road
Nr Oxford Street
WC2
KI 38
A24 PE P TL
A22 P
MMC HFS
03.33

In the appliance bay the three coloured bulbs lit up on the ceiling, red, green, yellow – the call was for all three appliances. Once again the crews donned their gear and climbed into their positions. Turk observed that the address on the call slip was wrong – he knew there was no Denmark Court off the Charing Cross Road. He reached for his radio.

[1] The crew that night were: Station Officer Martyn Manning, Sub-Officer Ron Morris, Paul Hale, John Bosse, Barry Dixie, Ron Beer, Geoff Gard, Jim Epson, John Gwynne, Chris Price, Steve Davies, Robby Fraser, Les Hemsley.

'M2FN from Alpha 242, priority, over!'

The calm voice of the radio operator at Wembley Fire Control answered reassuringly, 'Alpha 242, go ahead, over.'

'M2FN, from Alpha 242, from Station Officer Manning, request verification of address to Fire, Denmark Court. There is no such address on our ground, over.'

The control officer asked him to wait. On the appliances the crews finished rigging in their fire gear, while the drivers revved the engines in anticipation, like racers on the start line.

At this moment a further call was received, this time a 'running call'. This action operates the fire bells in a distinctive repeating fashion, making the call seem even more urgent. Ron Beer, the duty man, ran to the front of the station where he was informed by an anxious passer-by that a fire was in progress down an alleyway off Charing Cross Road, in Denmark Place. Without further hesitation, Turk shouted above the roar of the diesels, 'All appliances proceed to Denmark Place.'

Flames leap up from the roof of a fire in the heart of Soho; thick smoke obliterates the light.

Both fire engines parked in Charing Cross Road at the junction with Denmark Street. Turk glanced down Denmark Place and noticed a plastic sign on fire two-thirds of the way down the alley: it looked like another rubbish fire. As he walked towards it he came upon a man who was breathless and obviously injured. Until he spoke, Turk assumed that he had been involved in a fight. But then the man gasped, 'There are people in there . . . lots of them . . .'

Once Turk was alongside the building he could see smoke starting to escape from under the shuttered windows. Within a few seconds fierce flames began to lick from the windows at first and second floor levels. 'God Almighty, it's spreading fast!' he thought to himself. He shouted to his men, 'Get a hydrant set in and a jet, quickly.' He tried the front door, but it was locked and solid as a rock. (In fact it was of double thickness, steel-lined and fastened with concealed locks on the inside; it took the crew nearly four minutes to force it open.) The double front doors to the premises adjacent and directly underneath the fire were open and gave access to a large room containing various vendors' vehicles. One was overturned outside. A molten sign dripped fire down the face of the building from the first floor and the street was littered with burning rubbish and broken glass. 'Force this front door and get in there!' he commanded.

He then ran back to Charing Cross Road, where his driver, 'Singer' Hale, was laying out hose and preparing the pumps for action. 'Singer,' – Turk's voice was fiercely calm – 'make Pumps Four, Persons Reported.' Singer repeated the order and then ran to the front of the machine to send the priority assistance message.

It is one of the great traditions of Soho that whenever possible the first crews to arrive contain the fire without radioing for further assistance (a 'make up', whereby the number of pumping appliances is increased according to the additional number of men required for search and rescue purposes, the number of hose-lines required to extinguish the fire, the complexity of the building involved, the potential for the fire to spread unchecked and the number of actual pumps needed to supply the jets). This tradition of a reluctance to make up dates back to the days of intense rivalry between stations, but sometimes extra help is essential.

A 'make up' also describes how a serious fire progresses: there is a pre-set procedure whereby an officer normally asks for one additional pump at first. This indicates to the control room officers that a serious fire is in progress. In response to a 'Make Pumps Four' priority message, the control room then mobilizes a mobile control unit, a damage control tender (carrying equipment previously used by the London Salvage Corps) and an Assistant Divisional Officer. As soon as he is able, the officer in charge of the scene will send an informative message.

Turk returned to the alleyway to direct fire-fighting operations, and to his dismay found the pump crew still struggling to break open the front door. Shards of sparks and gleaming embers now rained down on them from above. For the crew the whole scene took on a surreal, 'Marie Celeste' quality. Many people were running off into the night, despite obvious injuries. The firemen were confronted with a severe fire in which they had been told people were trapped, yet they could hear no sound of anguished voices and could see no sign of people at the windows, as would normally have been expected. There was a very confined working area, it was densely smoke-logged and there was only one effective entry point into the building. They swore and cursed as they battled to open the door and from above the fire crackled and spat down on them.

Turk ordered Les Hemsley to run down Denmark Street, where he located a music shop and found that people were trapped behind the security grilles in the shop. Les could also see that the shop was very smoke-logged. An entry was effected and six people were rescued, most of them badly cut or burned. It appeared that the music shop was also involved in the fire, but when the crews ran out hose and entered the music shop on the upper floors they found that it was not actually ablaze. Glancing out the rear windows of the shop they saw the fire roaring out of the doors, windows and roof of the building behind in Denmark Place. Les Hemsley used this position as a vantage point from which to attack the fire and got his crew to work via the rear windows of the shop.

Turk then appeared from the Denmark Place side of the incident and got the turn-table ladder crew to pitch it over the roof of the music shop, in order to get a good view of the fire from above and to be ready to rescue any persons who might have made their way on to the roof. It could also have been used as a water tower to protect the adjacent premises which were fast becoming involved.

At approximately 0335 the Officer of the Watch in charge that night at Wembley Fire Control telephoned the Divisional Commander, Roy Baldwin, at his quarters above Paddington Fire Station, to inform him that a four pump fire was in progress on Soho's ground and that Assistant Divisional Officer Kennedy was proceeding. He responded immediately, although not required to by virtue of rank (normally a four pump fire only requires the attendance of an Assistant Divisional Officer):

But on this occasion something in my guts made me respond at once: I knew that Station Officer Manning would not make Pumps Four unless he had a good working job, as his 'Fours' were always worthy of a six pump attendance. En route to the fire at 0341 the radio crackled into life. 'M2FN from Alpha 242 priority, over.' 'Alpha 242 go ahead, over.' 'From Station Officer Manning, at Denmark Place WC2, make Pumps Six!'

On hearing this priority message I knew that my instinct had been right. I glanced knowingly at my driver, who responded by pushing his foot flat to the floor. I knew we had a good working fire in the best traditions of Soho, but never in my wildest dreams could I have anticipated the events that were to unfold during the next few hours. I arrived at 0348 and took charge of the fire two minutes later. On arrival I was met by ADO Kennedy, who informed me that he had a serious fire situation in progress, involving what was thought to be a large building, the front of the building being in Denmark Street and the rear in Denmark Place. The building had multiple occupancy and there were an unknown number of people involved.

He also informed me that propane heating cylinders were involved on the ground floor in amongst a collection of hot-dog stands. All around me the street was filled with foreign-looking people in great distress. I grabbed one of them and asked who they were and where they had come from, but was given only evasive replies. I took a policeman to one side and demanded full information regarding the premises involved and about all the screaming people in the street. He informed me that they had come out of two illegal clubs and that it had been known for as many as 150 people to be in the building at any one time. I made this information known to the crews about to enter the building, and instructed them to search for any injured persons.

At that time I decided that the six pump crews and turn-table ladder

Looking down on to the roof of the clubs, the full extent of the damage becomes clear. Firemen are standing on the remains of the metal fire escape in the doorway of the second floor club.

that were in attendance were enough to cope with the current fire situation. Several jets were already at work and the propane cylinders had been removed to safety and were no longer a hazard. I then accompanied ADO Kennedy and Station Officer Manning into number 18. It was now obvious that this was the main section of the fire and that the fire had spread from the two upper floors to the third floor of the adjoining office building. The fire crews were working hard but were encountering extreme heat.

Sub-Officer Ron Morris and his crew had by now gained access to the main staircase, and fought the fire successfully. The fireball had been so fierce that it had burned the roof away, and much of the heat vented out into the night sky above. Inside the building Ron, a real fireman's fireman, performed with his usual skill, using the minimum amount of water necessary to extinguish the fire and leaving the charred embers steaming but not drowned – the very art of good firemanship!

After about ten minutes Divisional Commander Baldwin made his way into the first floor bar, where Station Officer Manning said to him, 'I've found one charred body and I think there may be more, but it's hard to say, due to the severity of the fire still burning.' DC Baldwin then climbed the still smoking stairs up to the second floor and was again told that bodies had been found. As soon as the fire was extinguished in the top floor room, he did a quick head count of the bodies visible and at once dispatched a message to control indicating that eight bodies had been found. Recalling his actions at this moment in the fire, Fireman Steve Davies remembers the great moment when DC Baldwin turned to him on the second floor landing and uttered the immortal words, 'Fireman, don't let anybody in, *or out*, of this room.'

It was obvious to the fire officers that the fire had spread too rapidly to be an accident. Even though it had been burning for some fifteen minutes, the smell of a petroleum-based fuel was very evident around the foot of the stairs. A message was despatched for the brigade photographer to attend, and another requesting the police to begin their investigations as soon as possible, in view of the number of witnesses still in the area.

The fire was eventually extinguished with the use of three jets and breathing apparatus. The 'Stop' message was sent at 0518. By now the number of bodies had risen to eight on the first floor level and five on the second. At least fifty people had escaped alive; thirty of them were treated in various hospitals as a result of the fire, but many others skulked off into the night. Once the fire was out Mr Baldwin re-entered the building with a view to establishing the exact number of bodies involved, but this proved extremely difficult due to the way they were piled on top of each other and the confined space they had squeezed themselves into.

The final count was thirty-seven people killed, thirteen of them on the first floor, twenty-four on the second. The photographs of the bodies

The first floor of the club, looking towards the bar. There are seven bodies in this photograph.

show what an awesome task the crews of Soho had to face when it came to removing the remains.

DC Baldwin said later:

I have nothing but praise for the way in which the crews that made the initial attendance at the incident carried out their fire-fighting responsibilities. Though the building was not large, the spread of fire and the circumstances surrounding the fire taxed their resources to the limit. I appreciate that many of the firemen present had already had experience of dealing with dead bodies, but there were others who had no experience of handling dead persons before. Many of the tasks they performed during and after this fire were beyond the call of duty and I am proud and privileged to have served with such men. It is a difficult enough task even if the body is physically intact, but imagine if you can what it is like to be confronted by the remains of a human being burned beyond recognition, dismembered, stuck together in a bizarre manner, and so many. I asked the crews there who had seen burned bodies and who had not. I then took those who said they had not on a guided tour of the remains, for the sake of their education.

Opposite: Five charred bodies slumped in front of the bar.

Above: Divisional Commander Roy Baldwin talks to the press after the fire.

Working around dead bodies can have a profound effect upon the most experienced person. Burned bodies give off the most terrible smell, so repulsive that instincts demand one to escape: like rotting meat, burning hair and rancid vomit all rolled into one. Yet the men from Soho did not wish to be relieved from the fire until the job was done.

The next morning, when the fire was fully extinguished, the crews were finally changed. The fresh crews from the Red Watch now faced the horrific task of removing the charred bodies from the remains of the building. They worked the whole day, searching for bits of bodies in the smouldering ashes, a traumatic experience for all. One veteran fire officer told the press, 'I have never seen anything like it. People seem to have died on the spot without even having time to move an inch. Several of the bodies, terribly burnt, were simply slumped at tables where they had been sitting. The pile of seven bodies near one of the bars seemed to have fallen as they stood with drinks still in their hands.'

As the various parts of bodies were found, firemen attempted to label them with ordinary tie-on luggage labels. Wherever they were stuck together, or wherever a severed foot or arm was found, it was labelled, photographed and then removed.

It was decided that each body should be wrapped in a waterproof plastic sheet. The firemen from Soho did this work. Each body or piece of body had to be picked up by hand, and many had gone stiff with rigor mortis.

When they were placed in the sheets and lowered by line out of the windows, some pieces of the bodies broke away and either fell back into the black rubble or plunged gruesomely to the ground below. One body was being lowered when suddenly it tipped upside down and the entire contents of the stomach came slithering out on to the head of a policeman standing below. The next one slipped through and landed on the ground with a splat. This type of mishap has a strong effect on the men, as at all times they have the utmost sympathy for the victims and respect for the dead. The condition of both the bodies and the building in Denmark Place made it very difficult to carry out the job to be done with dignity.

DC Baldwin kept saying to Soho's men, 'Do you want to go back to the station for a breather?', but they chose to stay until about 1600 when the final body was removed, having started at 0900 that morning.

Blackened from head to foot and exhausted to the point of collapse, the Red Watch crew eventually returned to the station, had a good wash and a strong cup of tea, scrubbed and cleaned their equipment and as soon as possible had the appliances ready to respond to the next call Wembley Control would throw their way. 'No time for counselling or breaking down in 1980!' a fireman said later. (Nevertheless, a few of Soho's men had continued nightmares long after the Denmark Place disaster and ten years later, once a full counselling service had been established by Anne Wilmott and her skilled team of welfare officers, several men have needed to discuss the events of the Denmark Place fire.)[1]

On 25 August, nine days after the fire, Gypsy began recirculating in his old haunts. In the early hours of the morning, as he sat drinking a beer at 'Dave's' all-night club in Hanway Street, a man recognized him, walked to the pay-phone and quietly telephoned the police. A few minutes later he was arrested.

During the trial at the Old Bailey that followed, the defence QC suggested that Gypsy had been framed by gangland bosses, who had reputedly offered a £5,000 reward to anyone providing evidence that proved Gypsy's guilt. The taxi driver who had taken him to the petrol station was implicated, but he denied any knowledge of the reward.[2]

Gypsy Thompson was sentenced to life imprisonment for perpetrating the largest mass murder in modern British history. The fire also produced the most casualties of any fire in mainland Britain since the Second World War. The trial and fire were not given the press coverage they warranted, partly because the Yorkshire Ripper trial began on the same day as Gypsy's. Ask people in the street if they know what happened in Denmark Place and they will look at you with a blank face. The same is not true when you mention King's Cross underground station.

[1] The Red Watch crew who took over at Soho and spent most of that day removing bodies from the building were: Mick Woodard, Basil Brazier, Dave Smith, John Wilkinson, Bob Moulton, Dan Dare, Harry Willis, Tony Patterson, Don Clay and Ted Temple.

[2] Another fact that emerged during the trial was that prior to 16 August a betting shop in Charing Cross Road, very near Denmark Place, had had a quantity of flammable liquid poured through the letter box and ignited. Gypsy admitted that he had done this following an argument with the owner. He acknowledged that he wanted to get his own back 'gangster style'.

8 Death on the Underground and the King's Cross Fire

Literally millions of people pass through the West End of London each year, and many of them use the underground system. Soho Fire Station has responsibility for more underground stations than any other in the London Fire Brigade: Piccadilly Circus, Aldwych, Covent Garden, Temple, Oxford Circus, Tottenham Court Road, Charing Cross, Embankment and Leicester Square. Holborn, Bond Street, Green Park and Goodge Street stations are just off Soho's ground.

One of the most unpleasant tasks the London Fire Brigade is called upon to do is rescue people who have fallen, been pushed or thrown themselves into the path of an underground train. Most are people who have decided to kill themselves and usually they succeed.

Those who throw themselves in front of trains are known as 'jumpers': they usually jump in front of the train as it emerges from the tunnel. By the time the train has managed to stop it is a long way into the station, trapping the person under the middle of the train. The victims are either cut into sections which then drop below the lines or more often partially dismembered and entangled in the wheels of the train or impaled on the brake shoes and pinned to the underside of the train.

London Underground operate their own emergency engineering team, based in Cricklewood, who are responsible for the recovery of bodies, and normally it is they who deal with this task. However, if there is any possibility that a person may still be alive (which is not uncommon – people often live for fifteen or twenty minutes after having jumped) or if their removal is very complicated and protracted, the fire brigade are called. In such cases the fire crews work in close co-operation with the highly skilled London Underground team. Sometimes a distressed member of the public will telephone the fire brigade having witnessed a person falling under a train.

When the first fire crews arrive at the emergency they are often

Sub-Officer Terry Spindlow and Firemen Pete McCarlie, Ted Temple and Nobby Hall removing the body of a 'jumper', Charing Cross Station.

confronted by large numbers of irate passengers cursing the inconvenience. A few in the crowd wear ashen faces: they are the unfortunates who know why the station has been closed. The firemen walk briskly behind their London Transport guide, who shows them grimly to the offending platform. On the street above, sirens herald the arrival of additional fire appliances, ambulances and police cars. On the platform, each step nearer the train becomes heavier in anticipation. A shocked driver is comforted at one side. A station official has already located the body and summons the firemen forwards. – 'In there, mate!' – with the flick of a finger. Without a word the firemen begin the rescue, which usually involves the difficult and skilled lifting and manipulation of the train after the power to the electrified lines has been shut off.

The working conditions under a train are very restricted and claustrophobic: terrific heat builds up in the confined area and the rescue crews often have to operate in relays, the blood of the victim dripping on to them as they work.

On one occasion Soho's fire crew was called to rescue a man who had been running for a train at Goodge Street. As he jumped forwards, the doors closed. Part of his coat caught in the doors and he found himself holding on to the outside of the train as it moved off. To the horror of those left standing on the platform and those inside the carriage, he was

The twisted and dismembered remains of a man who has jumped under an underground train.

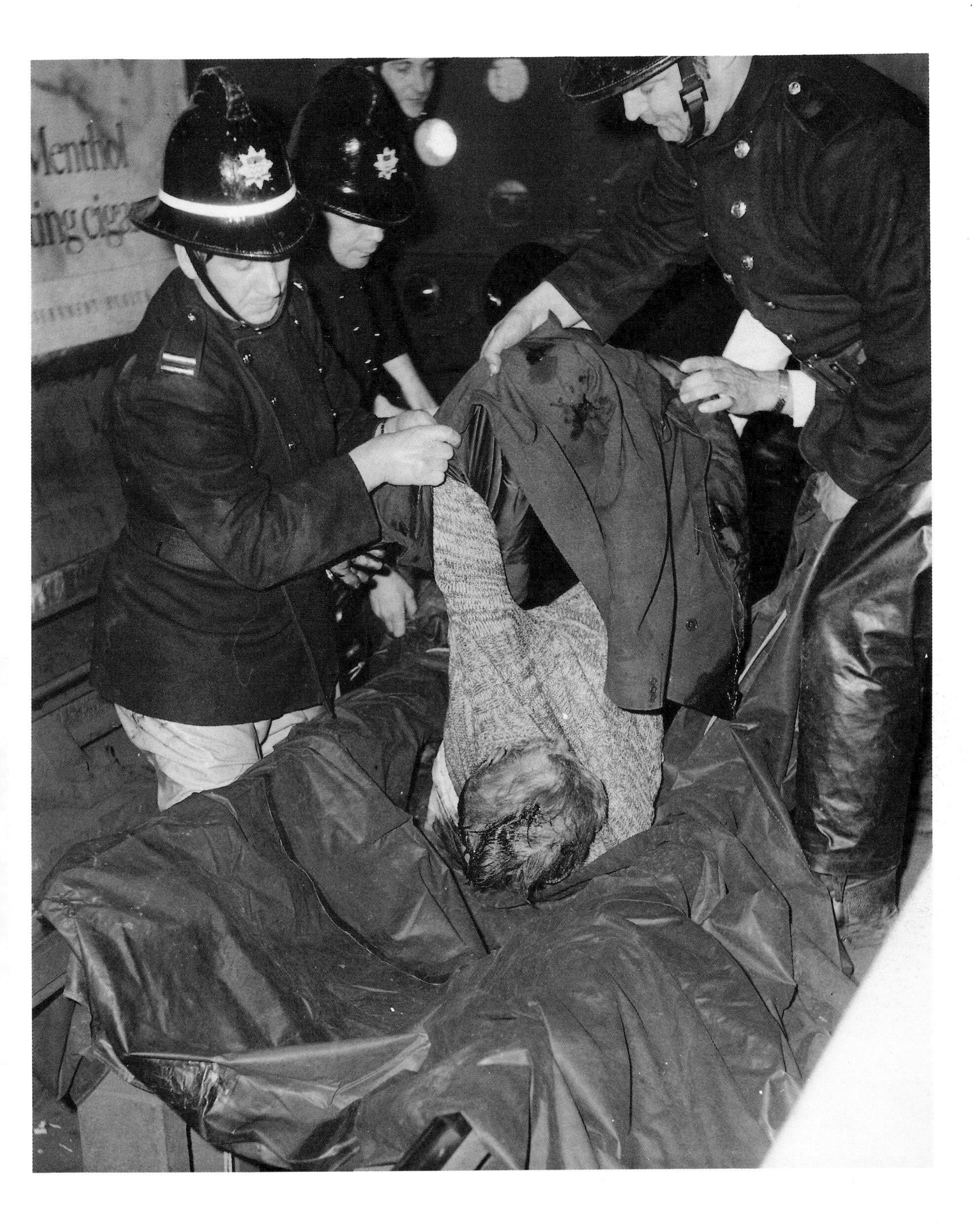
Menthol

dragged into the tunnel, and squashed between the train and the tunnel wall. Before anyone realized what was happening the train was twenty yards into the tunnel. At this point, a quick-witted passenger pushed the emergency stop button, alerting the driver to the emergency. Soho's crews crawled into the tunnel and found that the man was still conscious. They talked to him all the while he was being rescued, reassuring him and offering words of encouragement. 'Hold on, mate, don't worry, we'll have you out in a jiffy . . . you'll be all right.' With great speed and expertise the crew managed to lift the train in the tunnel to one side, enough to release the man. He was virtually unhurt.

Another man attempted to kill himself in front of a West End train. Roy Baldwin remembers the story with relish:

On arrival at the scene Soho's men found that he was trapped halfway along the underside of the train, wrapped around a wheel axle and wedged between the brake shoes. He was conscious and talked away while he was rescued. On peering down with my torch I could see that his left leg was cut clean off. He was losing a lot of blood and we realized that speed was of the essence. The crews were magnificent and released him rapidly.

A year later, almost to the day, Soho were called to another 'jumper' at the same station. On rescuing the man, they found that he had had his right leg cut off. As they were lifting his injured body on to the stretcher someone shouted, 'God almighty, his other leg is wood.' It was the same man who had lost his left leg the previous year![1]

Fires in basements or any underground building are always regarded as being particularly dangerous, characterized by confined space, lack of ventilation and light, and the build-up of high temperatures. Of all the possible underground fires that London's fire-fighters face, the most difficult are often in the underground transport system.

During the war the military authorities built large extensions off underground stations to use as air-raid shelters for soldiers passing through London during the Blitz. One of these was in Tottenham Court Road near Goodge Street Station. The tunnels were like station platforms but deep down in the ground, with access via a vertical shaft from ground level down a spiral staircase, similar to the emergency stairs you still see in most underground stations. These tunnels were lined with folding wooden bunks, as well as cooking equipment and basic washing facilities. After the war they were left down there, untouched. The one at Goodge Street was constructed on two levels. In 1954 Soho picked up a very serious working fire in two tunnels there. It was a hell of a job in which a number of

[1] Statistics for the deaths under trains at underground stations on Soho Fire Station's ground between 1977 and 1990:

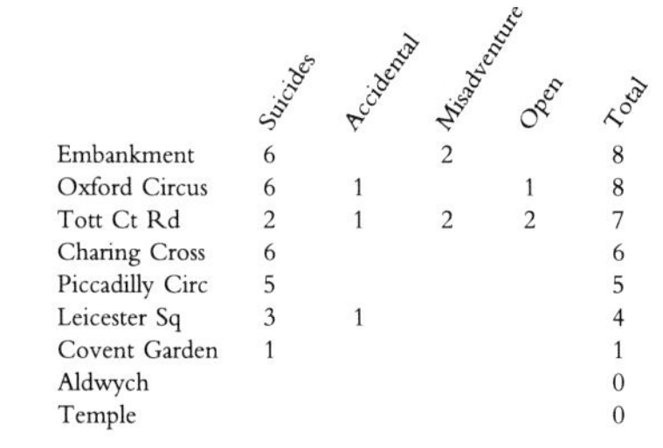

	Suicides	Accidental	Misadventure	Open	Total
Embankment	6		2		8
Oxford Circus	6	1		1	8
Tott Ct Rd	2	1	2	2	7
Charing Cross	6				6
Piccadilly Circ	5				5
Leicester Sq	3	1			4
Covent Garden	1				1
Aldwych					0
Temple					0

fireman were seriously injured. It had been caused by a reckless maintenance man discarding a cigarette.

In the late 1960s several firemen, including two of the most experienced officers in the brigade, were nearly killed at a small fire on the platform at Leicester Square. Dan Ivall was the Station Officer in charge, but because of the nature of the fire both Mr Cliff Colenutt, the Divisional Commander, and Mr Ernie Allday, the Assistant Chief Officer, came on. The fire was at the far end of a platform in some cable that unbeknown to the fire officers gave off a toxic acidic gas when in contact with water. The firemen actually fighting the fire were wearing BA, but the rest of the officers were not. Mr Allday and Station Officer Ivall stood talking at the exit end of the platform and after a brief discussion:

We decided to make our way back to street level as the fire had been successfully extinguished. As we reached the base of the escalator a train came into the station on an unaffected platform, causing a rush of air through the station, which had previously been very still. This pressure pushed the previously undetected gas along the platform and up the escalator we were climbing. It hit us like boiling oil being poured down your throat into your chest. Somehow we managed to get to the top and out into the open air, but it was touch and go.

Several other men were also overcome by the toxic fumes and all had to be taken to hospital. This fire too was believed to have been caused by a discarded match.

Smoking on underground trains has been banned for some years and fires in rail carriages are rare, but smoking in stations was not banned until 1985. As a result most fires in underground stations were caused, and still are, by the careless disposal of cigarettes or matches. Most vulnerable are the wooden escalators. Between 1956 and 1987 there were forty-six serious fires on escalators within the London Underground system and over two-thirds were attributed to smokers' materials.

On 11 June 1981 the Blue Watch was called to a fire at Covent Garden underground. Station Officer Tony Wilmott was in charge:

The fire was in a store under the spiral staircase. When my first crew went in, they were able to take a hose-line down, with some difficulty, but as they began to fight the fire, the actual staircase, being made of metal, began to get hotter and hotter until it was glowing at its base. The only way to get additional crews down to fight the fire was to order another attendance to Leicester Square, the adjacent station on the Piccadilly Line, and have the crews ride through the tunnel on a train, and approach the fire from the bottom of the spiral staircase.

The burned out remains of the escalators and ticket office area at Paddington Station, Christmas Eve 1944.

This method of approach was not considered operationally correct, but under the circumstances there was no alternative. It was a very unpleasant fire to fight, and required the crews of eight pumps to bring it under control. Against all odds, no one was hurt and the fire was contained to a relatively small area of the station.

Ten days later the Blue Watch picked up another underground job, this time at Goodge Street. On arrival they were greeted with a chaotic scene: distressed passengers pouring out on to the street, coughing and choking, and one casualty laid flat out on the ground, being given urgent resuscitation by a female passenger, quickly assisted by a fireman. Station Officer Wilmott was unable to find any London Transport staff to assist him in locating the fire, so he proceeded further into the station. At the top of the escalators he was shocked to see various passengers in a highly anxious state being assisted up the stairs by the station's staff. He promptly ordered his driver to 'Make Pumps Four, Persons Reported. Two crews to rig in breathing apparatus sets.' He was then informed that at least one train was on fire in the station and that people were possibly trapped below ground.

With this news the potential scale of crisis became apparent. Soho's crew flew into action, carrying out several rescues and effective first aid. Two minutes later Station Officer Wilmott was informed that another train was also at platform level and that it too probably contained passengers. He sent a further priority message, 'Make Pumps Eight.'

Below ground, the crews were experiencing great difficulty in locating the fire because of the thick smoke. Furthermore, they knew the power was still flowing through the tracks, which greatly restricted their freedom of movement. Despite several requests to London Transport staff to turn off the electric power to the tracks, this was not done, and the crews eventually located and fought the fire at great personal risk.

The fire was located in a semi-circular underground walkway between the north and southbound Northern Line tunnels, about forty yards into the main tunnel. It was eventually extinguished using the crews from twelve pumps: fifteen people had been removed to hospital for treatment, one of whom subsequently died, and a larger loss of life was avoided only by the valiant actions of the crews involved. This fire was plagued with communication problems both at the scene of the fire and between the London Fire Brigade control and London Transport. A variety of important recommendations for improvement were subsequently made.

The fire at Oxford Circus underground station was perhaps the most shocking fire of all prior to the King's Cross disaster. It occurred on the cold winter's night of 23 November 1984. At 2150 a passenger entered the operations room sited at the head of the escalators in the main booking hall and told a member of staff that there was a smell of smoke on the northbound Victoria Line platform. The staff member decided to investigate on his own, without telephoning the fire brigade. When he got down to the platform the fire was burning fiercely, but was isolated within an arched store running between the two platforms. As he attempted to summon help on an emergency telephone, conditions worsened rapidly, forcing him to retreat without alerting anyone else. Within a minute the whole platform was filled with smoke.

Wembley Fire Control received the first '999' call at 2202, twelve minutes after the fire had first been reported, and Soho's machines arrived four minutes later. The fire had thus gone unchecked for sixteen minutes before the first fire engines arrived. The White Watch was on duty, under the command of Station Officer Ray Chilton, a fire officer of great experience (decorated for his actions at the Worsley Hotel fire in 1974), and with a reputation for running a 'tight ship'.

The initial crews attending entered the main concourse of Oxford Circus station, which was virtually deserted. There was no smoke visible and as a result the firemen were not wearing breathing apparatus. Suddenly, with no prior warning, large volumes of smoke began punching up the escalators into the main ticket hall. The crew ran for their lives. Some made it to the street, the smoke chasing them like a wild tiger; others unwisely took refuge in the operations room at ticket concourse level.

Station Officer Chilton immediately made Pumps Four, Persons Reported, ordered every available man to rig in breathing apparatus and get jets to work. He was then informed that four members of London Regional Transport staff were trapped with three of his firemen in the operations room within the station. A rescue team were dispatched and all seven of the men were brought to safety. Seven minutes later Station Officer Chilton was informed that a train containing an unknown number of passengers was at the platform at the level of the fire. He immediately made Pumps Eight, and requested additional breathing apparatus. The fire crews experienced extreme difficulties as they encountered dense smoke and heat rushing up at them from deep below ground. It became apparent that the fire was spreading and many additional crews would be required. Eventually some crews located a train at the level of the fire and carried out a hazardous search for casualties, but the train proved to be empty. All the passengers had fled as soon as the train arrived in the station during the early stages of the fire.

Meanwhile two further trains full of passengers had been located between Oxford Circus and Green Park. The Assistant Chief Officer now in command of the fire-fighting operations made Pumps Twenty-Five, ordering some of the additional machines to proceed to Green Park to assist in the evacuation of passengers. As the full extent of the fire and the

One of the recently decorated platforms at Oxford Circus underground station, completely gutted when fire engulfed the station, 23 November 1984.

vast number of people involved became apparent pumps were made up to thirty, more than two hours after the original call. Two further trains containing passengers were located between Oxford Circus and Piccadilly Circus.

Fire crews finally located the main seat of the fire, which was ultimately beaten using four jets and a hundred breathing-apparatus sets. The whole of two platforms, their cross tunnels and a train were damaged by fire, heat and smoke, as were three miles of tunnel, escalators and terraced level concourse of offices and shops. Several trains had become stranded in tunnels around the fire once the power was shut off. The smoke and heat from the fire spread throughout these tunnels, causing great alarm to the passengers, who were stuck in total darkness. Eventually breathing apparatus crews who had walked through the maze of tunnels from Piccadilly Circus and other nearby stations reached the trapped passengers and led them to safety. There was a formal dance that night at the Café Royal Hotel in Regent Street and tearful reunions took place on the pavement outside Piccadilly Circus station as anxious party-goers found their missing partners, all dressed in ball gowns and black tie, and all with blackened faces like survivors from a mining disaster.

Nearly one thousand people were rescued from platforms and trains caught in the underground tunnels, of whom only fourteen required hospital treatment. By a miracle not one life was lost. But so serious was the anxiety of senior officers within the London Fire Brigade about the apparent two-tier procedure that underground staff were operating with regard to telephoning the fire brigade for help (whereby staff attempted to deal with all fire outbreaks themselves before summoning assistance) that in August 1985 the Chief Officer requested that DACO Kennedy write officially to the Operations Director of London Underground. His letter concluded 'I cannot urge too strongly . . . that clear instructions be given that on any suspicion of fire, the fire brigade be called without delay. This could save lives.'

The Oxford Circus fire was the greatest peacetime disaster (in terms of casualties) that never occurred in London – it was truly a damn near thing. As a result the Fire Safety Task Force was established and continued to hold meetings until May 1987. This task force was involved in the implementation of many of the actions recommended as a result of the fire, and closer collaboration with the London Fire Brigade ensued. The banning of smoking was extended in February 1985 to all areas beyond the ticket barrier at all stations underground.

Despite the recommendations, which themselves followed several previous requests to the Operational Director of London Underground from the 'A' Division Commander, Roy Baldwin, further fires followed in rapid succession. At 0800 on the morning of 22 December 1984, a fire was noted at the top of an escalator leading from the Piccadilly Line at Leicester Square station. The brigade were not called for six minutes, by which time the station was filled with smoke. In the meantime underground staff turned on the water-fog equipment in an attempt to control the fire. On arrival, the fire officer in charge was advised by the station inspector that they would need breathing apparatus and that a severe fire was in progress. The water-fog equipment had not put out the fire and it took crews twenty minutes to extinguish it. The station inspector and a woman police constable were taken to hospital suffering from smoke inhalation. A subsequent investigation concluded that the probable cause of the fire was a cigarette or match igniting rubbish under the skirting board. It was recommended that there be regular heavy-duty cleaning of all wooden escalators, that water-fog controls be located close to the machine room entrance and that staff be instructed always to use the water fog as soon as smoke or flames were observed.

A fire occurred the following month, on 25 January 1985, on the escalator at Green Park station. The fire was first noted at 1940, yet the brigade were not called until 1953. The first machine arrived seven minutes later, by which time the station was filled with dense smoke. Somehow all the passengers and staff were evacuated without injury and the fire was brought under control by fire-fighters in breathing apparatus using jets. It was later revealed that the water-fog system had not been actuated. The investigation that followed made a number of recommendations, including the proposal that plans of the station should be posted at station entrances for the benefit of the emergency services, that the controls for the water-fog equipment should be relocated outside the escalator machine room and that smoke detectors should be installed on escalators, in machine rooms and other areas of risk.

Despite the shocking number of fires that occurred after the Oxford Circus fire, many of the actions recommended had not been adequately implemented by the time of the King's Cross disaster. Had they been, events that night might have been very different.

In his conclusion to the investigation into the King's Cross underground fire, Desmond Fennell QC wrote:

For over a century London Underground has run an exceedingly safe railway system. It has a very good record and travel by the underground remains considerably safer than by almost every other form of transport. But London Underground, and its holding company London Regional Transport, had a blind spot – a belief that fires were inevitable, coupled

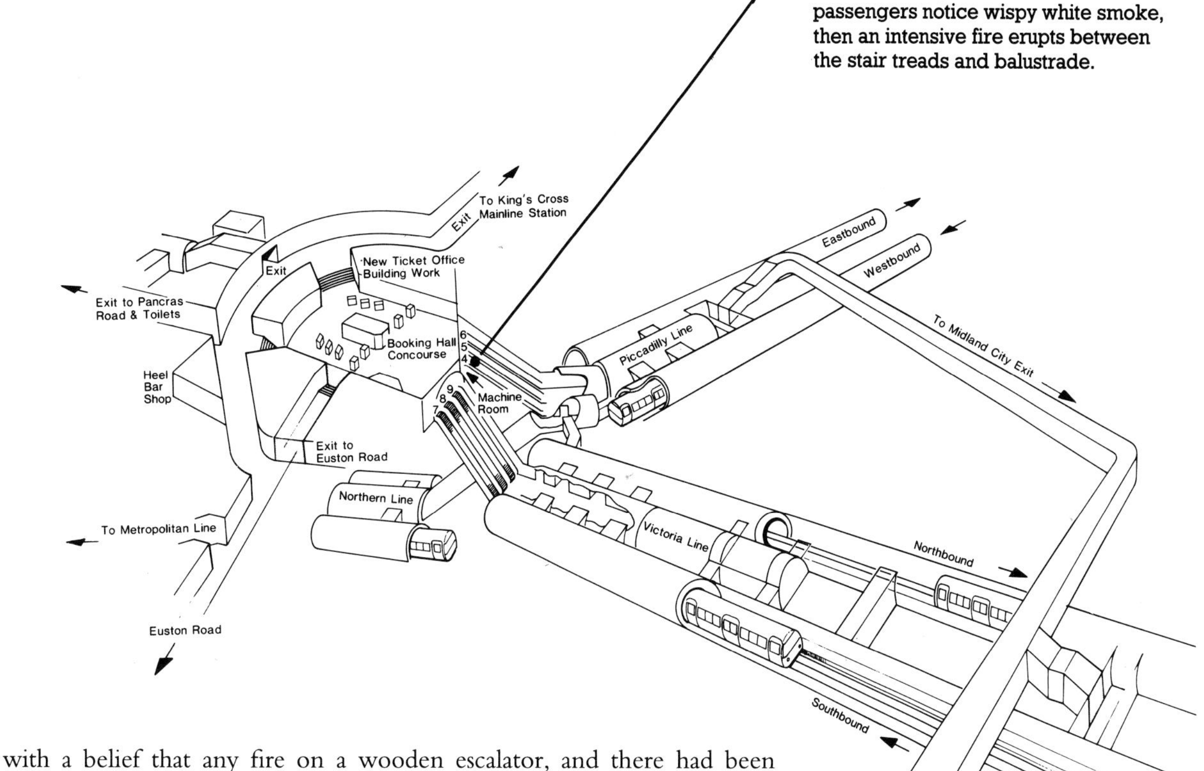

with a belief that any fire on a wooden escalator, and there had been many, would never develop in such a way as to endanger passengers. In my view that approach was seriously flawed, for it failed to recognize the unpredictability of fire, that most unpredictable of all hazards.

Sooner or later luck was going to run out. There had been too many near misses. Lessons had not been learned and London Transport had failed to implement the recommendations made as a result of a number of serious escalator fires. It was only a matter of time.[1]

Three-dimensional diagram of King's Cross underground station.

King's Cross Station is London's busiest underground, handling more than 73,000,000 passengers each year. During the peak of the evening rush hour (between 1600 and 1830) as many as 100,000 passengers use the station. Five underground lines converge at this point: the Piccadilly, Northern, Victoria, Metropolitan and Circle.

At approximately 1929, Mr Philip Squire was travelling up the Number 4 escalator, leading from the Piccadilly Line platforms, when he noticed white smoke coming from underneath the stair treads. As soon as he reached the ticket hall he reported the fire to Mr Derek Newman, the booking hall clerk. A minute later Abdeslam Karmoun saw smoke and fire two-thirds of the way up the same escalator, pressed the emergency stop diamond at the top and shouted down to people behind him, 'Get off, get

[1] Crews from many fire stations fought the fire at King's Cross. In the early stages, machines arrived from Soho, Clerkenwell, Manchester Square and Westminster. The details of the fire are complicated and a complete record of the actions taken by the many valiant crews who were there that night would require a book in itself. Therefore this account is mainly concerned with the actions of Soho's men.

off, it's on fire.' At 1935 Mr Myatt was halfway up the same escalator when he also noticed wisps of smoke coming from both sides and smelled burning rubber. Suddenly he saw a more alarming three-foot line of sparks coming from the right-hand side, from a gap between the steps and bordering below the hand-rail. As he arrived at the top, so did PCs Terry Bebbington and Ken Kerbey, having seen Mr Karmoun press the emergency stop button.

When Mrs Capes arrived at the foot of the same escalator it had already stopped moving. About halfway up she noticed grey smoke at ankle height curling up the side panel to the right-hand side. Then she saw a fierce fire between the tread and skirting panel. She ran back down and waited at the bottom for a minute, obviously wondering what to do next. Then she decided to go up via Number 6 escalator. At the top she found a London Transport policeman. She took him at once to the top of Number 4 escalator. From here they could see thick dark smoke coming from where the rubber hand-rail returns, near the floor. No flames were visible anywhere.

PC Kerbey attempted to radio for assistance from within the station, but his radio would not function below ground and he ran to street level via the Euston Road exit. He radioed his urgent message, 'This is an active message. We have a fire on the Piccadilly Line escalator at King's Cross.

The Soho Red Watch. Back row, left to right: Wally Slade, Ted Temple, Mick Woodard, David Priestman, Dave Smith, Martin Fittall, Bob Moulton, Rodney Cordell, Steve Bell, Stewart Button, Martin Powell. Front row: John Edgar, Mark Blakeman, Paul Hale, Roger Kendall, Vernon Trefry, Jeff Hale, John Wilkinson.

Request further assistance and fire service immediately.' The first 999 call to the fire brigade was received by Wembley Fire Control from the British Transport Police at 1936.

ESCALATOR ALIGHT
KINGS CROSS LT TRANSPORT
PICCADILLY LINE
PANCRAS ROAD NW1
40 K2 62 A23 CAM CRR HAD TSR
C27 PL
A24 PL TL
A24 P BY RT
A22 P
N AREA FCU NE AREA ACU[1]

Station Officer Colin Townsley, known as 'Tonka', rode in charge of the pump ladder. He was a tall, handsome man, always immaculately turned out. He was extremely fit and had been nicknamed after the 'indestructible toy' by the Red Watch crew at Soho who held him in high esteem. John Wilkinson drove the pump ladder, while Bob Moulton, John Edgar, Stewart Button and Martin Powell sat in the back. The traffic was heavy as always. They turned right out of the station into Shaftesbury Avenue and first left up Charing Cross Road. They had a good start on the pump, which was out and about on the ground doing hydrant tests, but it still took them six long minutes to reach their destination. The pump ladder from Soho was the first appliance to arrive.

King's Cross is on Euston Fire Station's ground, not Soho's, but the machines from Euston were already out on another call. Euston Fire Station is less than two minutes 'on the bell' from King's Cross. One can only speculate as to what might have happened if they had been available.

Leading Fireman Roger Kendall on the Soho pump was driven by Martin Fittall. In the back were Dave Smith, Rodney Cordell, Steven 'Dingers' Bell and David Priestman.

Paul Hale drove the 'ladders' with Sub-Officer Vernon Trefry in charge. Both these machines arrived eight minutes after the initial call.

When Soho's pump ladder arrived, it parked outside the St Pancras entrance in Pancras Road. Firemen Wilkinson and Powell stayed with the machine, whilst Station Officer Townsley and Firemen Moulton, Button and Edgar went down into the station to investigate. They saw no sign of fire or smoke and everything appeared normal. They then moved as a group to the top of Number 4 escalator. From here they could see that the fire was burning about twenty feet down the stairs. The flames were bright

[1] 40 K2 62 is the map route and reference. CAM indicates the London Borough of Camden, CRR HAD is the Central Risk Register code as explained in Chapter 5, and TSR stands for Temporary Special Risk, indicating that the premises had some temporary additional hazard about which information was stored at Wembley Fire Control, C27 is Clerkenwell Fire Station, A22 is Manchester Square Fire Station. The North Area Forward Control Unit and the North East Area Control Unit are also both despatched as part of the pre-determined attendance.

orange, burning well and leaping four feet in the air. Strangely, there was no smoke at all. Colin turned to his crew and said in his ever calm voice, 'Go and get your breathing apparatus on, and get a jet.' As they were hurrying out of the station, Bob Moulton stopped an underground employee and asked him where the hose-reel was, but the man shrugged his shoulders and pointed at the fire. Some passengers stood transfixed staring at the fire and others began to make for the exits.

By now the pump from Clerkenwell Fire Station had arrived. On entering the subway, the crew could smell smoke and their officer in charge, Sub-Officer Bell, ordered his crew back out to get rigged in breathing-apparatus sets. He proceeded into the station with one of his crew and met Colin Townsley at the top of the escalator. After a moment's discussion Colin walked down the escalator furthest away from the one burning. He shouted up at Sub-Officer Bell, 'We're going to need sets. I'm going to make Pumps Four, Persons Reported.' Bell then ran past the fire calling out, 'You go up and take it from there, Guv, and I'll go down and stop the people from coming up.'

At the same moment, Station Officer Pete Osborne from Manchester Square Fire Station had entered the concourse with his crew. Quickly assessing the fire, he too ordered his crew back out to rig in BA and to get a jet set in. He walked over to Colin and they talked briefly. Three of Clerkenwell's crew had now come back down into the station concourse in breathing apparatus. They stood between the ticket barrier and the top of the escalator. Colin approached their leading fireman, Flanagan, saying, 'Go back and make Pumps Four, Persons Reported.' People were still coming up the Victoria Line escalator into the concourse. Some of them looked quite alarmed. There was little smoke, but a strong smell of burning and the noise of the fire crackling on the stairs.

As Flanagan walked towards his crew, Pete Osborne turned away from Colin, distracted by a commotion at the top of the Victoria Line escalators. As he walked over to investigate, there was a sudden crash, like falling debris. He looked up. To his horror the ceiling was disappearing under a curtain of dense black smoke filled with flame. Osborne was an officer with some nineteen years experience: instinct took over and he ran straight down the Victoria Line escalator nearest him. He shouted at the other passengers to run back down to the bottom and assisted a badly burned man down the stairs, where he cooled his injuries with a water fire extinguisher. Despite being much closer to the top of the burning escalator, Colin was able to run towards the nearest exit. As he ran, he shouted at passengers to evacuate the station. A woman near to him was struck down by the wall of heat and without hesitation he went back for her, and assisted her along the exit tunnel.

They were but yards from the exit when he too was overcome by the heat and smoke and collapsed.

In a matter of two or three seconds the fire had burst out from the mouth of the escalators and into the main concourse, in a fireman's worst nightmare, a flash-over: there is a sudden increase in temperature as gases reach a critical temperature and then explode, propelling a wall of searing, scalding heat that pushes forward in front of the chasing smoke and flame. Clerkenwell's crew did not even have time to start up their breathing-apparatus sets, they just ran for their lives! The concourse was plunged into complete darkness and the air filled with terrible screams as everyone caught in the hellish blast was cut down, stripped of clothes and skin.

Somehow Clerkenwell's crew managed to make it out of the Pancras Road exit, dragging as many people as they could with them. At this moment Soho's turn-table ladder and pump pulled up. Fireman 'Dingers' Bell was on the pump:

We were in Trafalgar Square doing hydrants when we got the shout over the radio. The TL [turn-table ladder] had also been ordered and as we turned into Charing Cross Road it was about twenty yards in front of us, also proceeding on the bell.

The traffic was quite heavy and we did our usual high speed dodgem dance in and out of the jam. Just before we crossed Tottenham Court Road, the ladder's back plate fell off into the road with a flash of sparks, but they did not notice and carried on oblivious. We stopped, jumped out, retrieved it and went on our way again. That only took a couple of seconds. We did not realize how vital those seconds would become. We arrived at King's Cross a couple of minutes later. The rest of the crew jumped off the machine and were starting to walk towards the TL when I suddenly heard Martin Fittall's voice saying, 'Hang on a minute, I think we might have a job here.' I stuck my head out of the window and glanced towards the entrance stairs and noticed a few grey wisps of smoke coming out into the street.

Next moment, smoke began to pour out: dark, black, thick smoke. Ugly black. Pumping out like it was coming out of a steam-train funnel. The staircase was filled with smoke and the entrance at its base just disappeared. The crew on the pavement could hear screaming coming from below. Without waiting for orders they immediately jumped on to the pump and began to rig in BA. They climbed over the pedestrian barriers and straight down into the now solid wall of smoke. As they reached the bottom step a couple of people stumbled past them. Somehow they had made it to fresh air.

Martin Fittall set in a jet from the hydrant to the pump and the crew took that back down the stairs, waited just a couple of seconds for the 'water on' and as soon as it came started to inch forwards. At the bottom of the stairs they turned right into the corridor and the heat hit them like they'd never felt it before, their ears starting to burn instantly. They fell flat against the floor where it is usually cooler but the heat was just as intense. They began to move slightly away from the cover of the water jet, feeling round on the floor for any casualties, but everywhere they tried to go they got burnt, their fingers and hands going numb in the heat.

Suddenly a man just appeared out of the smoke and fell down on top of them. As they couldn't see more than three inches in front of their faces, his appearance came as quite a shock. Rodney Cordell and Dave Smith dragged him to the bottom of the stairs where 'Singer' Hale joined them and together they took him up to the street. Martin Fittall tried to look after him, but his clothes were burned off and his skull had split open in the pressure of the heat.

Dingers, Vernon and David were finding progress very hard going:

The spray wouldn't work on the nozzle properly, so we had half a main jet and only half a water curtain. Slowly we edged forwards along the twenty yards of tunnel towards the main concourse, and every now and again it

Aerial plan of the main ticket concourse.

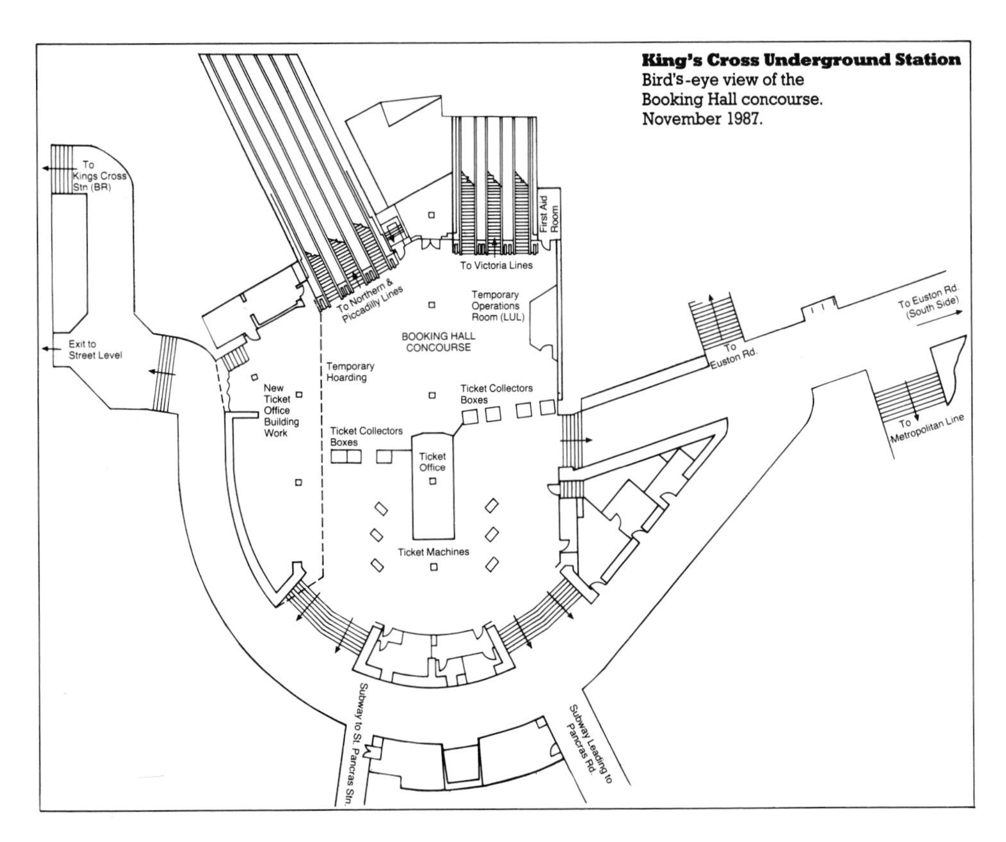

would suddenly go as clear as day. We realized the trains were still running. The smoke was being sucked back down into the fire as they entered and left the platforms. The sight that greeted us was completely unprecedented: the jet of water was simply turning to steam not more than ten feet after leaving the nozzle. The spray curtain was also turning to steam and as a result provided us with very little cover. Every time a train came into the station a great rush of air would fly up the escalator and feed the fire afresh. The waves of heat washed over us like a flame-thrower.

On the surface there was considerable confusion, caused partly by the large number of badly injured casualties emerging on to the pavement and partly by the fact that the four senior fire officers present had all disappeared in the fire. More and more firemen and officers were arriving but there was something of a breakdown in communication. Large volumes of smoke shrouded the whole area around King's Cross as ambulances began to arrive in growing numbers. Leading Fireman Kendall, having assisted several distressed members of the public from the bottom of the stairs, found himself in charge of the breathing apparatus tally board, a much-maligned task vital to the safety of the men below ground. Normally the BA control man will have a maximum of twelve tallies on his board. That night Kendall found himself with forty – an immense and unenviable responsibility.

While this was going on, Moulton, Buttons and Edgar had re-entered the Pancras Road entrance. They experienced intense heat and found it extremely hard work even to negotiate the first flight of steps. At the foot of the second flight they found the body of a woman which they removed to street level. She was so badly burned that they could not even prise her mouth open to administer the kiss of life. They left her in the care of firemen on the surface and went back into the subway. Under cover of the jet Bob Moulton edged forwards a few yards:

The men on the jet were hitting my back with it so I was able to move forwards. I had gone about five or six feet when I suddenly felt a helmet in my hands. I shone my torch on to it, and taking my masked face close, could make out the black stripe of a Station Officer's helmet. Then I found his head and felt that he was not wearing breathing apparatus. I immediately knelt over him and dragged him towards the stairs. It was bloody hot, and very hard work. I knocked his helmet off whilst doing this . . . I went straight up the stairs, grabbed some help and we brought him out.

'God, there's no pulse,' someone said desperately. Without another word we ripped his tunic open and started cardiac massage. I began mouth to mouth. Someone ran over and handed me a brook airway tube. After a short while an ambulance crew and a doctor arrived and they took over. I knew it was bad then, but there was no time to stop. We had other things to do.

Meanwhile Dingers, Vernon and David were making further progress into the fire. As they reached a small flight of stairs which led down into the main concourse, the jet of water appeared to reach not even as far as the ticket office before it turned into steam and then disappeared completely. At this point Rodney Cordell and Dave Smith rejoined them, having come back down with a hose-reel. From the stairs they could see the glow of the fire all around and could roughly make out the main ticket office, which was well alight, and the top of the escalator belching flame. As Dave and Rodney came up behind them with their jet on full spray, the actual metal of the nozzle became so hot that they could hardly bear to hold it. As soon as their spray hit the front crew on the back they were off into the main concourse. At the height of the fire the temperature in the concourse was about one thousand degrees Celsius. Rodney and Dave set off down the right-hand side of the stairs while Dingers, Vernon and David pushed on towards the escalators with a view to fighting the fire there.

Everything seemed to be burning. The fire travelled across the roof in ripples, and it looked as if one ceiling tile was setting fire to the one next to it. Molten plastic dripped on to them from the ceiling; it landed in the steaming water, forming strange shapes on the floor. Quite soon they made direct contact with the fire in the wooden ticket offices and managed to put them out.

Vernon checked the gauges on the breathing-apparatus sets. David Priestman's was very low and he was sent out. The others were OK and carried on. As they moved closer to Rodney and Dave they came across the first body. It was very badly charred and hardly recognizable as human. In the darkness they could make out that it was small, probably a child:

In amongst the black mess I suddenly saw the sinister spectre of a hand and arm from the elbow up, sticking straight up like a swan's neck. I was crawling forwards and it was only a couple of inches away from my face when I saw it clearly. As we had first gone down the steps in the station a woman was screaming, 'My child is still in there.' Everything was happening so fast it went in one ear and out the other. But when I saw the small body, it hit me. I thought, 'Oh, dear God, that's her kid there.'

The noise of the fire was frightening: metal panels banging and buckling in the blistering heat, tiles cracking and flying off the walls, hoardings splitting. By now they had got a long way in, but still couldn't get to the

Fire crews remove the badly burned body of one of the victims.

actual seat of the fire. Everything was burning. And in this strange red light they could now see bodies everywhere. The crews had been fully committed for about fifteen minutes. They were completely exhausted. That sort of heat draws energy from the body all the time. Sweating pints, their bodies demanded a lot of air, using up their supplies fast, hardly giving them the strength to hold up the jet.

Suddenly Dingers' warning whistle started to go, telling him that his air supply was running low:

I knew I must get out, but there was nothing to lash the jet on to. It was acting as a cover for the hose-reel crew and if I'd taken the jet away the heat would have rebounded on to them. I kept talking to myself all the time: 'Once the whistle goes off . . . you've always been told you've got about four or five minutes, but then in this heat maybe I've got only two minutes . . . and now I can't remember how long the whistle's been blowing.'

Dave Smith, who had gone out to change his cylinder, came back down, this time following the main hose to Dingers. As soon as he touched him on the shoulder Dingers began to make his exit. He started to crawl back along the hose, when all of a sudden there was an horrendous bang as one of the ticket machines buckled and he pressed himself against the hot ground. In doing so he had let go of the hose, his lifeline to the outside world. It was gone. He reached here, then there, a bit further that way, then this way, talking to himself all the time, 'Find your way out . . . do the old test for heat. Use the backs of your hands in the air . . .' But that didn't work because his hands were burned wherever he tried putting them. He had to do something, so he set off blindly into the oven. Then bang! Out of the blue something hit him on the head. He looked up and, taking his torch in hand, moved right up to see what it was, his face mask almost touching it. It was the end of the brass hand-rail on the same small set of stairs. He was right at the entrance into the corridor and just hadn't seen it. Dingers crawled along the corridor, clawing his way towards the shaft of light up ahead. As he reached the bottom of the main stairs the whistle stopped. He had completely run out of air:

I walked up two steps, looked up, saw a blur of steps that went on forever, thought, 'I'm not going to make it', took another step. Then I felt something warm running down my legs. I was pissing myself. I took another step, then collapsed. My mind was clear as anything, but my body just stopped working. Martin and Singer picked me up and dragged me to the top of the stairs, took my set off and sat me down, leaning against a railing. Someone chucked an ice-cold bucket of water over my head, which left me steaming in the night air. I kept thinking to myself, 'Now I've made a right fool of myself, giving out and pissing myself. I'm going to get a right ripping from the boys for this', when all of a sudden there were other blokes emerging from the stairs, collapsing all around me. Going down like lemmings they were. Everywhere. Dave Smith collapsed next to me and I helped him off with his set. When he'd recovered a bit I said to him, 'Dave, something funny's happened to me.' As I said it I must have glanced down at my crotch because he said, 'Oh what, have you pissed yourself?' 'Yeah.' 'Oh thank God, somebody else has done it as well.' He sounded really relieved.

Soon after Dingers made his exit, Malcolm 'Otis' Redding, a big ox of a fellow from Westminster, emerged on to the stairs in some distress. Moments earlier Vernon had got into trouble and Redding rescued him when he had collapsed down in the concourse, but in doing so had completely done himself in. Otis was helped up the stairs by Roger Kendall and others. He sat down to lean against a pillar and next second he was flat out on his back. Several of Soho's men ran forwards to assist in his revival and the television crews which had now arrived in their hordes captured this little drama on film to be broadcast within hours to every corner of the globe.

When the fire was finally put out, the fire crews had one more task to perform: the removal the bodies. Owing to the complexity and size of the fire few of the men from Soho had even spoken to each other: none of them yet knew that their Guvnor was dead. Uncomplaining, too tired to talk, they lined the bodies up behind the hoarding in front of the station, away from the intrusive glare of the television cameras. A large van arrived from the mortuary, and many bodies were loaded into it, although some had already gone in ambulances.

The dead were carried in salvage sheets. Said one of the crew:

I was helping carry one body and I had hold of its leg, which was over my shoulder. I was so tired that I lost my footing a couple of times. As we crossed the road I tripped over. The sheet fell off and all I could see was half a leg. The other half was completely burned away, straight down to the bone, whereas on the other side, it was completely intact. You could see the edge of the skin, the muscles and tendons, everything. Like a slice of a cake. His trousers had been burned off but he was still wearing half a sock and half a leather shoe. Looking at it, it didn't effect me that much, but the smell. God, that's a smell you don't forget.

Most of the bodies had become almost inhuman. They were so badly burned, and so completely dehydrated, that they had shrunk to a third of their normal size: 'small black shrivelled heaps of carbon'.

A fireman walked over to where some of Soho's crew were standing together and said calmly, 'Oh, mate, Tonka's dead.' Martin Fittall looked shocked and disbelieving. The unknown fireman disappeared back into the mêlée before he could be questioned further. Suddenly Assistant Chief Officer Jim McMillan called them over towards him. He said ominously, 'Get all of Soho's crew together. I want a word.'

Hiding his own grief, and knowing the importance of the way in which he broke the news to them, he would follow the now legendary style of the great and sadly missed Charles Clisby whose job it had previously been to break the news of dead firemen to their friends and families. Very solemnly he took them to one side. As they gathered together they were all talking nervously. Someone said that Bob had pulled the Guvnor out and given him mouth to mouth. Someone else had seen Neil Johnson from 'the Square' (Manchester Square Fire Station) giving the Guvnor cardiac massage, and thought that he hadn't looked good, so they knew it was bad. ACO McMillan spoke very calmly, but in a loud and clear voice, 'Station Officer Townsley was removed by ambulance from this fire and has subsequently died. We will arrange for you boys to get away, back home as soon as possible.'

Everyone was stunned into silence. Then, as the numbing shock began to wear off, first one, then, as they saw each other, more began to let the tears flow down their grimy faces. Great long white streaks appeared across their blackened cheeks. In their grief they got their stuff together and drove home to Soho, the Guvnor's seat empty, shouting its terrible symbolic news at them, news they did not want to believe. Wilks kept glancing at the empty seat beside him, not believing, just not believing . . .

Back at the station the atmosphere was heavy. Leading Fireman Roger Kendall found himself in charge, and as such acted up to the rank of Sub-Officer. The pressure on him was terrible. He made the difficult decision to get the men working again as soon as possible. After tear-filled tea and silent showers the men set about re-stowing the appliances with equipment, replacing their breathing-apparatus set cylinders and cleaning their blackened, stained uniforms and helmets. Within a couple of hours the machines were back on the run. Almost immediately the bells went down again, this time calling them to an automatic fire alarm actuating in a building in Oxford Street. The stress of this first call cannot be over-emphasized, but the cathartic result was surely healing. In this case, as so frustratingly often with these calls, it proved to be a false alarm.

On returning to the station they were besieged by the press. Reporters shouted through the letter box, banged on the windows and even attempted to climb over the fire station's back yard wall. The ugly side of Fleet Street screamed at them, 'Let's see his locker . . . let us take a picture of his uniform hanging up . . . give us an exclusive, it'll be worth your while.' They were not allowed in but the men were sick at heart.

Anne Wilmott and her team from the Welfare Department mobilized all their resources to care for the injured men and their families, as well as Colin's family. Over the following months many men who were there that night were to ask for counselling. Others have left the job for ever, the long-term effects of this traumatic fire being too much for them to bear. The events of that night filled their sleep with nightmares. The screaming that the initial crews standing on the pavements during the flash-over heard haunted them for months:

It was the most terrible screaming. You could feel their terror and actually hear the pain as their chests burned in the tidal wave of heat that first chased them along the corridor and then caught up with them. After about ten seconds of this terrible screaming, silence. Total silence, like after a bomb has gone off.

The next morning the world awoke to news of the disaster. Mark 'Carrots'

Blakeman, Wally Slade, Mick Woodard and Ted Temple had not been riding the previous night, but their grief was no less for that:

It was a self-inflicted feeling: on the one hand being glad that you weren't there in case you were one of the unfortunates and on the other hand a feeling of overpowering guilt. The blokes had all aged a lifetime. It was really hard for us to come back to see them afterwards.

That evening the men were solemn as they lined up for roll call. Roger Kendall stood smartly to attention and ordered his men to do the same. Then they stood in silence, biting lips, swallowing hard, fighting back the tears. As the names were read out and each man answered, the overwhelming sadness in many of them bubbled over again. Colin Townsley had been the father of the watch and now his 'children' were grieving for him . . .

Soho Fire Station was flooded with floral and written tributes from all over the world. The firemen of Clermont Ferrand in France presented the station with a slate plaque bearing their brigade badge and motto, 'Sauver ou Périr', and the words, 'Save or Die – it meant something for you.'

On Friday 27 November 1987, central London came to a standstill. Soho's Blue, Green and White Watches lined Shaftesbury Avenue as the flower-laden turn-table ladder carrying the coffin turned out of the station, behind the pump and pump ladder, where Colin's seat lay empty. Several of the Red Watch crew had worked all through the previous night arranging the flowers on the ladders. The cortège moved through the throng of West End traffic, following the same route they had taken the previous week. At King's Cross they paused for a minute as a thousand-strong guard of honour stood in splendid silence.

From there they drove to St Paul's, Covent Garden. Dave Smith, David Priestman, Rodney Cordell, Roger Kendall, Vernon Trefry and Mark Blakeman carried their Guvnor with pride, and more tears were shed. The people of Soho lined the streets and many of them cried openly though they had never known him. At the service a poem was read out:

> No need to ask them questions,
> their faces tell it all,
> They came back riding four,
> they were five before the call.
> Silent rage controls them,
> with an aching, numbing pain,
> As the anger wells within them,
> their tears they can't contain.

Was it just another shout,
 a nothing job gone wrong?
They know that's not the case,
 the menace there was strong.
Death placed his hand among them,
 and stole away their mate,
A man who strived for others,
 whose caring . . . sealed his fate.

But when the grieving's done,
 and the bitterness is spent,
They'll speak again of Colin,
 of what he did and meant,
And each will write an epitaph,
 to the Guv'nor they all knew
Who gave his life in a job he loved,
 a fireman through and through.

One of the hundreds of spectacular floral tributes in the form of a Station Officer's helmet.

Fire officers from all over the United Kingdom line the route in tribute to Colin Townsley as his body arrives at Lewisham Crematorium. White chrysanthemums form the word 'Guv': a poignant parting gift from the men of Soho Red Watch; a small card is inscribed 'Never forgotten, Guv.'

A few days later the following letter from Station Officer Ray Chilton, White Watch Soho, was published in the *Independent*:

On the surface, Colin was larger than life – a hard man who did not find it easy to tolerate those who did not reach his high expectations or the high standards he set in all the aspects of his life. But scratch that surface and you would find a gentle, sensitive, creative man, who adored his wife and his daughters and had a great affection for his men, Soho fire station, cycling, music and literature.

Months later the Honours and Awards Board of the London Fire and Civil Defence Authority granted many awards as a result of the King's Cross fire. Their number included Colin Townsley, who 'following the flash-over . . . immediately directed the evacuation of the concourse and delayed his escape to assist a badly burned young woman along the St Pancras subway . . . Station Officer Townsley was a fit and strong man and would almost certainly have reached safety but for his courageous and unselfish act in helping another.' Subsequently Colin Townsley was posthumously awarded the George Medal for Gallantry by Her Majesty The Queen. The award was accepted on his behalf by his parents.

It seems strange and sad that, perhaps as in war, it is often those who particularly deserve awards who do not receive them. Three of Soho's crew that night were conspicuous by their absence from the honours list. Said one, 'You just want to feel that you are valued . . . that you have some worth . . . that you are important to the brigade.' Colin Townsley's men knew that he expected them all to act beyond the call of duty without question, and that fateful night at King's Cross, all of Soho's men did just that.

LOOKING TO THE FUTURE

The London Fire Brigade has changed beyond all recognition in the last few years: in equipment, size, organization and personnel. Equipment in particular has developed dramatically. The brigade has a fleet of new vehicles 'on the run', including new pumping appliances, large mobile control units, and the very fast, turbo-charged Camiva turn-table ladders (one of which is based at Soho). Heavy and light rescue units have replaced the old emergency rescue tenders – the heavy rescue unit carries a vast array of specialized equipment, including heavy lifting, pulling, cutting and forced entry equipment and also extended duration (one hour, lightweight) breathing-apparatus sets. The light rescue units carry technologically advanced cutting and spreading rescue equipment, extended duration breathing apparatus and line rescue equipment. All its crews are also trained paramedics, making them better equipped to save lives.

Traditionally, although all fire-fighters are trained in first aid, advanced casualty care has been the responsibility of the London Ambulance Service. In reality it is usually the fire-fighters who arrive first. Some argue that there is a strong case for upgrading the first-aid training of a large proportion of fire-fighters to paramedic levels, without wishing in any way to undermine the work of the London Ambulance Service. In the United States most firemen carry defibrillators and intubation devices as part of their standard equipment.

While the debate continues, the London Ambulance Service has introduced a new paramedic service to London, including advanced paramedics who proceed to emergencies on motorcycles fitted with sirens and blue flashing lights, similar to police motorcycles, and carrying a comprehensive array of sophisticated first-aid equipment. The use of motorcycles enables the rider to negotiate heavily congested roads rapidly and improve response times significantly.

Other new equipment being tested includes computerized breathing-apparatus sets that will enable BA control staff to monitor the cylinder's pressure, the temperature of the room in which the wearer is operating, his location and even the volume of air being consumed by the fire-fighter each minute he is in a burning building. It also incorporates a radio communication system and a display board inside the face mask to indicate to the user when their cylinder contents have reached half working duration, and again when their air pressure is low. The breathing apparatus cylinders presently used by London's firemen are heavy and cumbersome: it is hoped that within the next few years a lighter-weight and more compact cylinder will be introduced.

Each year the brigade answers more and more calls. Soho, the busiest station in the London Fire Brigade, responded to 8,210 emergency calls in 1990. Consequently the quantity of radio traffic between appliances on the fire ground and Command and Mobilising Centre increases proportionally. It is planned that within the next few years each appliance will be fitted with a computer-linked pre-coded message-sender called the 'button box'. At present, when the officer in charge of an incident wishes to send the 'stop' message, informing control that the incident is over, he does so verbally using a mobile radio telephone. The 'stop' messages are all number-coded: Code 1 indicates the incident was a fire which resulted in the damage of property, Code 6 indicates a malicious false alarm and Code 7 a special service such as a person shut in a lift. The button boxes will mean that the officer in charge simply presses Button 1 if he wishes to send a Code 1 stop. This information will be recorded on the main mobilizing computer without any precious radio time having been wasted.

The vast deluge of office paperwork that swamps the junior officers in the fire station should also soon be replaced by a computerized system whereby each station has its own computer console networked to a mainframe.

When the radio operator at Command and Mobilising Centre needs to speak to the driver of an appliance attending an incident, the driver is often unable to hear the radio, if he is working away from the cab of the appliance. The fitting of loudspeakers on the outside of fire appliances to relay radio messages is also under consideration. The speed of progress, as always, will depend on financial constraints.

In the name of progress, and because of an inexplicable obsession on the part of the authorities with standardization of equipment, London's fire-fighters have in the last few years lost some of their traditional equipment, which, although antiquated in design, demanding in training hours and expensive to build and maintain, offered the most effective and arguably most important rescue tools of their trade.

The spectacular rescue of ten people from a building in St James's (1924), described in Chapter Three, was one of many notable events in the remarkable history of the hook ladder in London. Between 1955 and 1982 in central London alone, eighty-one people were rescued via hook ladders and hundreds of others were rescued via escape and turn-table ladders, assisted by firemen on hook ladders. Yet in 1984 the ladders were withdrawn from use.

During the last week that hook ladders were 'on the run', Steve King and Malcolm Burns from Kensington Fire Station attended a serious hotel fire in Cromwell Road and carried out a brilliant hook-ladder rescue at the back of the premises which would not have been possible with any other ladder. All eleven Divisional Commanders of the London Fire Brigade argued strongly for its retention in London. In researching this book the author has interviewed eight Assistant Chief Officers, five Senior Divisional Officers, four other Divisional Officers, ten Assistant Divisional Officers, dozens of Station Officers, and many other junior officers and fire-fighters (who are, after all, the front-line troops who have to use the equipment). Every single one of these men believes that removing the hook ladder from London was a mistake and that it should be brought back into service. It is beyond comprehension that this powerful lobby should be ignored.

Soho probably needs hook ladders more than any other ground in London: architecturally the West End is so complex and crowded; many of its buildings have wells in their centres, buildings within buildings, passageways that open into large courtyards, etc. No ladder carried on the machines today gives the firemen the same ability of access and the freedom to scale the outside of a building to any floor as the hook ladder.

Roy Baldwin, who retired as an Assistant Chief Officer, argues the same is true, to a lesser extent, for the scaling ladder, also now withdrawn from service:

It was a wooden ladder, tapered at the top, widening at the base, used to fit on to the next one, using bracket fittings. Each section was six foot six inches long. I remember a serious fire in the British Home Stores in Oxford Street. The fire was going in the middle of the building in a chute which ran from top to bottom. I ordered Soho's crew to use the scaling ladders because nothing else would have allowed them to get to it, as it was so narrow. We had six on top of each other. Quickly we got a hose-reel up there and stopped it. We could have lost the building otherwise.

Another Assistant Chief Officer stated: 'If you have a piece of equipment that has saved a life even once, where no other piece of equipment would

John Spreadbury, photographed in the gym at Soho Fire Station, wears the gold medal he won in the power lifting competition at the first World Fire-fighters' Games, held in New Zealand in 1990. 'Spreaders' also beat all the heavyweight category records, which makes him officially the strongest fireman in the world!

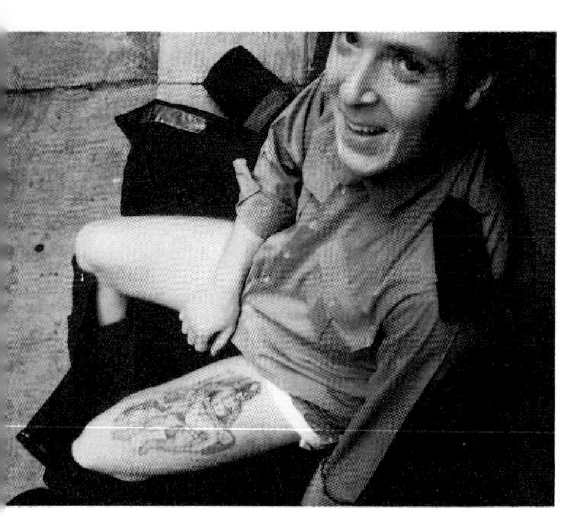

Several hundred pounds poorer and after hours of extremely painful visits to a Soho tattooist, Pete Blackman is now the proud owner of the famous painting by Charles Vigor, *Saved*. The fireman's number, on the breast pocket of his tunic, has been changed from the original to read A24, Soho's station number.

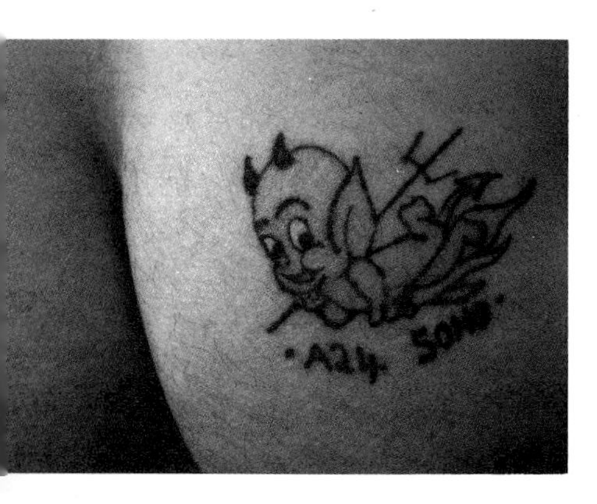

The love of Soho Fire Station knows no bounds.

have worked, then it should be kept on the run, whatever the cost.'

Traditionally the brigade's principal rescue ladder was the fifty-foot wheeled wooden escape. It was a fast, flexible, stable, strong, manoeuvrable, versatile ladder developed, designed and modified over 150 years of use in central London. It has recently been replaced by an alternative aluminium ladder known as the 1-3-5 (it is 13.5 metres – about forty-four feet – in length). This ladder, although very fine, is not the most suitable for rescue work in central London. Ideally the 1-3-5 ladder should be carried on the pump appliances, while in the heart of London, pump escapes should be re-introduced. The loss of escapes was seen as a great setback by all the men and officers of Soho. Other brigades in the British Isles which have residential streets of similar design to central London still have escape ladders on the run – two such brigades are the Isle of Man and Hertfordshire Fire and Rescue Service (Watford). The Isle of Wight and East Sussex Fire Brigades withdrew their last pump escapes from service in 1991 due to financial pressures, but East Sussex admits that if it had the resources, the escapes would still be on the run, as they are still the 'best ladder for the job'. Unfortunately, increasingly mean fiscal policies have forced fire brigades to cut vital equipment countrywide.

Some have argued that both the escape ladders and the hook ladders were withdrawn from service as they required too much physical effort to enable female fire-fighters to pass through training school. Others blame the Fire Brigades' Union for using its influential lobby to demand excessive standards of health and safety in relation to all aspects of practical fire-fighting and equipment used. Of course health and safety are vital components in all working environments, but Soho's fire-fighters see the ever-increasing pressures put upon them and senior officers to observe almost oppressive demands for safety as impractical in the real world of fire-fighting.

Nevertheless a vast improvement in much equipment has occurred in the last few years, including a dramatic change of fire-fighting uniform from the Victorian-style heavy cloth tunics, silver buttons, yellow plastic leggings, cork helmets, axe and belt, to a new long coat and matching trouser made of Nomex (a man-made fire-proofed fibre similar to a thick cotton), new leather boots with steel toe-caps, and advanced Kevlar helmets with face visors (the Kevlar surface covers a polycarbonate base similar to a motorcycle helmet). Sadly the personal axes were withdrawn, too, although there is talk of re-introducing this vital piece of equipment in the near future.

The third major rescue ladder in the London Fire Brigade is the aerial appliance: either in the form of a hydraulic platform (HP) or a hundred-foot turn-table ladder (TL). A TL or HP is ordered to most fire calls in central London. Often they are not required at the incident, but they are invaluable for high level rescue. Unfortunately, again due to lack of resources, there are only sixteen aerial appliances in the London Fire Brigade. In most capital cities of the world, pumps outnumber aerials by three to one. In London it is twelve to one.

Soho Fire Station in 1991 has a pump, pump ladder and turn-table ladder. The TL is very busy because it attends so many fire calls on other stations' grounds. As a result of such continual use it often breaks down, putting even more pressure on the remaining TLs in the central London area. A Station Officer at Soho recently related the following nightmare:

The ladders are on the run but as usual are out on another call, this time in Hammersmith, to an automatic fire alarm actuating at the Charing Cross Hospital. While the ladders are out, the bells go down calling us to a fire at the Strand Palace Hotel. Our pair [the pump and pump ladder] respond. The nearest available TL is dispatched by control from Kensington Fire Station. It is the rush hour. On arrival, we are confronted by a major fire involving the first floor of the hotel. Smoke is pouring out of every floor above. People are hanging out of windows at the second, third and fourth floor level, all screaming for rescue. Others have made their way on to the roof and are also imploring for help. I know that it will take Kensington's TL at least twelve minutes to arrive. Then someone tells me that there are people requiring rescue at the back of the building on the fourth floor. I have no escape ladder, no hook ladders and no aerial ladder.

This frightening scenario is not unrealistic but sadly, until there is a major disaster where lives are lost due to the lack of appropriate rescue ladders, it is unlikely that the necessary money will be found to re-introduce the hook and escape ladders to central London fire stations.

Over the last few years there has been a shift of emphasis in the role of front line fire-fighters: traditionally their duty was first to the saving of life, and second, to the saving of property. Now their first duty is to life but great emphasis is also laid on the prevention of fires. The safety of firemen and firewomen entering a burning building to fight the fire is paramount, more important even that the actual extinction of the fire. As a result, senior officers in charge of incidents tend to withdraw crews from fires if any serious risk to their safety is apparent: the cavalier style of tackling a fire in hand-to-hand close combat so beloved of the London Fire Brigade, taking the jet of water to the very heart of the fire, is now seen as reckless and irresponsible. As a result some fires do 'get away' (spread out of

Wardour Street, May 1991: wearing the very latest fire-fighting tunics, leggings and helmets, Sub-Officer Pat Millea and Firewoman Carmen Curren, Red Watch, prepare to tackle an electrical fire in a swish film-editing suite with BCF (one of a group of halogenated hydrocarbon extinguishers which are of particular use with highly sensitive electrical equipment whose delicate circuitry would be damaged by powder or carbon dioxide extinguishers).

The blue device labelled 'Big Ben' on their shoulders is the latest distress signal unit. On 'starting up' (switching on the breathing apparatus), the DSU is also turned on. This device senses movement and activates a shrill electronic alarm if it records 'no movement' for longer than twenty seconds. Thus if a fireman collapses or becomes trapped, the alarm is raised very quickly. In addition, the breathing apparatus is fitted with a warning whistle which activates when the wearer has between five and ten minutes of air supply left in his cylinder. However, a fireman following operational procedure should continually monitor his pressure gauge and be out in the safety of fresh air long before the whistle goes off.

control), but fewer firemen are reported injured. Whilst in an ideal world fire prevention will become so effective that fires will not occur, in reality this is about as likely as local authorities giving fire brigades the financial resources they actually need to provide the optimum service, or the government funding a major national review of fire-service provision in this country, which is long overdue. There will always be fires, and putting them out will always be a very dangerous business.

The contentious debate of whether women should be allowed to fight fires alongside men has raged for more than a hundred years. In 1870 the *Fireman* magazine printed the following article:

One of the employments in which woman is about to compete with man is that of service in the fire brigade. The first step in this direction has been taken in America, where, at one of the colleges for young ladies, a brigade of firewomen, or rather of firegirls is, it is stated, in course of organization. The notion is to accustom the fair members of the brigade to exhibit the self-possession so necessary in moments of emergency, such as the outbreak of fire, and at the same time to teach them how to act with promptitude on such occasions. The brigade is to be divided into groups, each group with its own captain and fire engine. The pupils are also to be taught the management of a steam fire engine until they are as efficient at the pump as at a piano. Whether the women will generally take to this occupation of extinguishing fire is doubtful, although there is no question that they often exhibit a courage and sang froid when houses they inhabit are on fire which is not always displayed by those who were in former days termed their 'natural protectors'. On the other hand, they sometimes go off in hysterics, while vile and stupid man, to do him justice, is not so frequently subject to this distressing and inconvenient affection.

The *Fireman* was written for an almost exclusively male readership, but the magazine maintained a gallant position of modest support for female firemen. A few years later it reported significant fire-fighting events taking place much closer to home: at Cambridge University:

The lady students of Girton College having lately had cause for alarm over the overheating of one of the stove pipes in the building, resolved on forming themselves into a fire brigade . . . They can work the engines and execute the manoeuvres with the greatest precision and when Captain Shaw was called upon to witness their drill, he was fain to compliment them heartily on their proficiency.

BIG BEN

The New York Fire Department first recruited women into the department in 1978 but all of the 170 women who took the physical failed. The test included dragging charged hose, climbing walls, raising ladders, climbing and descending ladders, simulated forcible entry using a sledgehammer, tunnel crawl and dummy drag, all within four minutes. In 1988 sixteen women out of a total of eighty-seven achieved sufficient scores to enable them to gain access to the NYFD, though most of them had attended a pre-test training programme.

The first female fire-fighters in London past through the training school at Southwark in 1983. One of them was posted to Soho White Watch, which resulted in a most unhappy episode for the station. Most firemen had never previously worked with women: just prior to her arrival the men were given no specific advice or training on the implications of having a woman fire-fighter on the watch. 'Treat her as an equal,' people said, and they did, taking this to mean 'treat her like a man' – with all the usual rowdy jokes and tricks that firemen always play on one another. The subsequent disciplinary hearings were harsh: an example had to be set to the rest of the brigade. Female fire-fighters were here to stay, and woe betide any fireman who overstepped the mark with the new arrivals. One of the firemen involved was sacked, several were demoted and most were posted to other fire stations. The firewoman concerned left employment with the brigade soon afterwards.

A modern London Fire Brigade crew must be multi-talented, extremely fit, strong, quick-witted, fast-thinking and above all, calm: 'headless chickens are not helpful.' But in addition each member of the team will also have a special talent. As Station Officer Steve Short put it:

Different skills go with different physical shapes: if there's a heavy door that needs kicking in, I'll instinctively look for the 'big lad', but if the only point of access to a building is through a tiny window at third floor level, I'm going to choose a small member of the team, and the chances are that the smallest person in the crew will be a woman, and in certain circumstances will be absolutely invaluable. It's easy for firemen who have still to work directly alongside women to criticize them in the job, but it would do them good to take a long hard look at the men they work with: some of them will be overweight, unfit, skivers or skates. Let the men put their own house in order before they start universally writing off women fire-fighters as losers.

Carmen Curren, fire-fighter, joined the Red Watch at Soho in 1990:

Christmas 1990.

Ever since I was five years old when I saw a fire engine in our street I wanted to be in the fire brigade. My first professional job was as a carpenter working for a council. I was one of the only girls who worked as a 'chippy' and used to get a variety of reactions from the blokes. I had a good, busy job, but all the time the fire brigade was in the back of my mind. It's true that people think that the fire brigade is still a man's world. It's also true that in general men are stronger than women, but it does not follow that all men are strong and all women weak. To do this job well you have to have enough strength, not be superman.

When I learned that the age limit for entry into the London Fire Brigade was thirty, that pushed me into action. I was just twenty-nine and kept pestering the recruitment centre again and again until eventually I got on to the ladder and started the very long selection process. Written tests in comprehension, dictation and mathematics, a full medical and eye-sight test, and a strength test too. The test used to be to carry a man at the run across the yard and back, and to hand wind a simulated fifty-foot escape ladder to full extension and then back down again, but this has been changed to leg and back pull tests and grip tests, all done on machines.

Then there is also a step test, where you have to step up and down on to a box to measure your aerobic capacity. The requirements are exactly the same for men and women in all aspects, including height (five foot six). If you pass all these initial tests and an interview, you then get assigned to a squad at Southwark Training Centre. The squad is usually about fourteen, including one or sometimes two girls.

The full training course is twenty weeks of intensive practical and theoretical teaching. The workload is so heavy that if you miss three days training you are back-squadded [demoted to a 'younger' squad]. The day begins at 8.30 a.m. with a parade and continues non-stop until 4.30 p.m. The training school is run like a military academy, with great attention paid to drill, marching, turn out and discipline.

When a recruit comes to a station he or she is on probation and will not be expected to be active on the fire ground without the guiding hand of an experienced senior hand or junior officer. Many brigades still require their recruits to undergo realistic training sessions in 'heat and smoke fire houses' where crates of timber are set alight in mock houses, the rooms heated to a high temperature and filled with thick smoke, to give potential fire-fighters a taste of the shocking atmospheres in which they are often expected to work. The London Fire Brigade does not require its recruits to have even seen a flicker of flame nor a breath of hot air when they go to a fire station, which is perhaps to be regretted.

However, the public of Soho react well to female fire-fighters. At a call to an automatic fire alarm actuating in one of the nurses' halls of residence, a group of girls nudged each other as Carmen accompanied her male Red Watch colleagues into the building. 'Is it a girl?' one asked curiously. Carmen flashed a broad smile and a proud nod in reply. The nurses beamed with approval. 'That's right grand!' shouted one, obviously thrilled, in a broad Yorkshire brogue.

A new-style London fire-fighter is emerging in the nineties – professional, fit, ambitious, health-conscious and enthusiastic. The crews of the Nineties comprise both men and women.

'It's brilliant at Soho,' says Carmen, 'like having fourteen big brothers. Yes, they do give me a lot of stick, but that's right. They give it to each other, and I give it back, too. Behind all the teasing and joking I know that they are looking out for me on the fire ground, and that I can count on them if things get nasty. I love Soho because it's right in the middle of the action!'

In 1991 Soho has two of the forty-three female fire-fighters in London. There are a total of 6,500 uniformed firemen in the brigade (a ratio of 150:1).

POSTSCRIPT

I have had the privilege of riding at Soho Fire Station for ten years. In that time I have attended more than 15,000 emergency calls. I never cease to be amazed by the extraordinary skills of Soho's fire-fighters that I have witnessed first hand. My quest to photograph the front-line action has been blighted by a jinx under which I have missed over 250 serious working fires on Soho's ground since 1982. However, I have been to many others and have seen with my own eyes men, and now women, go into the doorway of hell itself and back. Despite all the changes, there is one aspect of the job which remains as thrilling as it was in the days of the galloping, snorting, glistening, horse-drawn, steam fire engines: the explosively dynamic, sensory, exciting experience of the ride to a roaring fire . . .

On 15 August 1989 the Green Watch prepare for roll call. It is a warm summer's evening, and thousands of commuters swarm along the pavements of Shaftesbury Avenue past the closed doors of the fire station. Behind the red doors Station Officer Chris Staynings calls Green Watch to attention. As the roll is called, Phil Taylor interrupts, saying, 'Guvnor, I can smell burning.'

'So can I,' says John Bosse.

At 1801 Wembley Fire Control receive the first of fifty-three '999' calls to 'Fire, 15 Lower Regent Street, W1'. A minute later the bells go down. Each man, wherever he stands, glances up at the coloured light-bulbs on the ceiling of the appliance bay. Red! (for the pump ladder) and Green! (for the pump). (Had the turn-table ladder been on the run the yellow bulb would also have flashed to life.)

'Pump ladder and pump!' cries the dutyman as he comes out of the watchroom, shouting the address of the fire.

The feeling of exhilaration is almost overwhelming; the anticipation of the charge into battle euphoric. The automatic station doors swirl open, like a giant yawning to reveal its blood-stained teeth, and with a roar, the engines growl into life, spewing clouds of diesel exhaust into the air, and then the dragon spits its two bursts of fire into Shaftesbury Avenue. In the street, wisps of smoke waft over the roof-tops. The race begins: beams of blue light dance around the street like frenzied tropical insects in the warm London air. Cars and buses peel away in the machines' path, and cabbies wave them on. Station Officer Staynings gasps in disbelief: the street ahead looks as if a thick fog has descended on it, blotting out the sky.

'We've got a job!' the men shout to each other enthusiastically, struggling into their breathing-apparatus sets as the machines veer and lunge through the congested traffic. On, on down the avenue, passers-by covering their ears trying to shut out the air-blasting two-tone horns and the new American-style whooping and wailing electronic sirens. Sharp left into Windmill Street, surging and lurching between the cars, over the red lights at Coventry Street, confused foreign pedestrians leaping out of the path of the unstoppable locomotion. A suicidal cyclist slaloms in front of the pump ladder, then falls off into the gutter just as it seems inevitable he will be pulverized under the wheels of the scarlet beast. On, on down Haymarket. A quick glance right: a tantalizing glimpse of flame. Then over on two wheels into Pall Mall at the bottom.

'Oh God, Dan, not the old wall of death!' shouts Chris Staynings as the convoy swerves around the next corner and into Lower Regent Street. A crowd of pointing people filling the street turn as one face to greet the appliances' arrival. Even as the engines are stopping the crews dismount, adjust sets and prepare to take on the beast that roars and growls angrily above them. In seconds the drivers are 'setting the hydrant in', ready to receive the lengths of rolled hose which the rest of the crew throw vigorously along the road surface.

In minutes Soho's two crews are tackling the roaring fire at very close quarters. They know there will be LPG and oxyacetylene cylinders involved in the fire as the building is undergoing a refurbishment. The fire has got a good hold on the upper floors. It will take time and additional crews to beat it into submission, but for the moment they are stopping it spreading any further. Then a senior officer arrives and decides very quickly that the danger from exploding cylinders and the extent and spread of the fire is such that Soho's crews must be withdrawn to safer positions. As a result the fire spreads rapidly, and eventually twenty pumps are needed to extinguish it. Soho's crew are disappointed that they have not been allowed to run the gauntlet.

'It's a sign of the times,' says one, 'but it was still a bloody good fire and under the circumstances we did a damn good job.'

Technology and change may continue apace, but despite the inevitable teething troubles that the introduction of new equipment brings, and the murmurs of discontent about the excessive numbers of officers in the brigade, the loss of hook and escape ladders, the power of the union and the health and safety lobby, the crews at the sharp end love the work they do with a passion that becomes a way of life. Those who try to turn the career of fire-fighting into a job are doomed to failure: life in the fire service is a vocation, a calling, a labour of love. The fire-fighters of Soho perform a supreme task that demands the very highest standards of physical and psychological strength and courage. There is no other job, and no other place quite like Soho, in the whole world.

The bottom line is this: whoever you are and whatever you do, when your house is on fire, or your arm is trapped in a meat grinder, or you find yourself in a dark broken-down lift in an empty office block, and you also happen to be in Soho, then take comfort in your moment of distress when you hear the blaring sirens of the approaching fire engines. Rescue is about to arrive: the most wonderful, unassuming, unpretentious, brave, skilled team of reluctant heroes and heroines: the real life cavalry. Thank God!

Glossary

ACO: Assistant Chief Officer, the third most senior rank in the London Fire Brigade.
ACU: Area Control Unit – large mobile control room. Attends six pump fires and above. Manned by staff officers, the ACU provides the focal point for command and control at large incidents.
Additional: term used to mobilize an extra appliance to a fire call (usually at the discretion of the officer in charge of mobilizing control, and often on receipt of 'additional' calls to an incident).
ADO: Assistant Divisional Officer; the rank above station officer.
Aerial: fire appliance with the capability of reaching the upper floors of buildings i.e. hydraulic platforms and turn-table ladders.
AFA: Automatic Fire Alarm – umbrella term to describe a variety of automatic heat and smoke detectors that, when actuated, set off a fire alarm.
AFS: Auxiliary Fire Service, established in 1937 to recruit and train additional firemen and women in preparation for the outbreak of the Second World War.
'All the Lot': watchroom man's cry to indicate that all three fire appliances are to turn out.
Arson: wilfully or maliciously setting fire to property.
BA: see Breathing Apparatus.
Backdraft: see Flash-over.
Bardic: small powerful yellow plastic torch worn by all fire-fighters.
Beano: fireman's lingo for a day or weekend trip away with the boys.
Biff: fireman's expression for a crash involving a fire engine.
Branch: the business end of an attack hose line.
Breathing Apparatus: collective term for variety of different compressed air cylinders worn with face masks to enable fire-fighters to work in thick smoke or poisonous environments.
Brown Bread: rhyming slang for dead.
Buck: name given to probationary fire-fighters (also junior buck).
Camel: camel's hump – rhyming slang for pump.
Charged Line: a hose line with water coursing through it.
Chinatown: that area of Soho immediately behind the fire station, roughly contained within the borders of Charing Cross Road, Lisle Street, Wardour Street and Shaftesbury Avenue.
Chopper: axe.
Claret: blood.
CMC: Command and Mobilising Centre, Lambeth, where all the '999' emergency fire calls are processed. This high-tech computerized complex is staffed by control-room staff with the huge responsibility of mobilizing and commanding all the appliances in the London Fire Brigade.
Coachman: fireman responsible for the horses at the station and for driving the engine to the fire.
Conflagration: a fire of such tremendous intensity that water jets have virtually no effect on it.
DCU: Damage Control Unit.
DO: Divisional Officer – senior-ranking fire officer.
Door Jolly: Soho expression for standing out at the front of the station watching the world go by on a Saturday night.
Dry Riser (Dry Rising Main): a pipe built into high-rise buildings with outlets on each floor to enable fire-fighters to attach charged hose at ground floor level, then carry hose to any floor in the building and attach their hoses as if they were being set in to a ground level hydrant. Some buildings also have wet risers, which are similar except that they are already charged with water.
DSU: Distress Signal Unit, worn by breathing-apparatus crews and used to alert other fire-fighters in case of becoming trapped or collapsing unconscious.
Engineer: fireman responsible for laying the fire under the steam engine, lighting the fire before turning out to get up the steam and manning the pump on reaching the fire.
FCU: Forward Control Unit, usually a Range Rover, driven by staff officers to the scene of four pump fires and above, and other specialized incidents where additional command and control officers are required.
Fire: the rapid oxidation of gaseous, liquid or solid fuel in atmospheric air.
Fire Alarm Post: sand-bagged street-based fire station manned by auxiliary fire personnel.
Flame: the visible heat-producing element of fire.
Flash-over: a sudden intensifying of the fire when an inrush of oxygen occurs, perhaps because of a window or door being opened. An often deadly threat, also known as backdraft.
Goer: a fire that is burning strongly or 'going well'.
Guv or **Guvnor:** term of address used by firemen for the Station Officer.
'Hi Hi Hi!': the fireman's shout to clear the way for the charging steam appliance.
Hook Ladder: wooden ladder, thirteen feet in length, fitted with a hinged tensile steel hook, and used to scale buildings by hooking the ladder over the windowsill of the floor above.
Horsed Escape: fifty-foot wheeled ladder transported to fires on specially adapted cart pulled by two horses.
HP: Hydraulic Platform fire appliance with elevating beams and a cage or platform at the head, from which firemen can carry out fire-fighting operations or into which people can be transferred and carried to the ground.
Hydrant: water supply point from mains found in every street, usually under the surface of the pavement and

marked on an adjacent wall by a yellow descriptive plaque.
Hydraulics: the science of the flow of liquids, and particularly the fireman's science of moving water from one point to another.
Jet: hose line.
Job: working fire in LFB lingo.
'Jumper': a person who is threatening to jump, or who has already jumped, off a building or under a train.
Knock it down, or **back:** hit and extinguish the fire with water.
Knock off: turn off the water supply.
Leggings: over-trousers, previously of plastic, now made of Nomex fire-resistant man-made fibre.
Line: rope, used for a variety of purposes including lowering people from upper floors, rescues, hauling hose aloft, tying back beams, tying jets in position, etc.
Lobby: Victorian name for the watchroom.
London Fire Brigade: formed in 1904 to replace the MFB.
London Fire Engine Establishment: the first municipal fire brigade, formed in 1833.
London Salvage Corps: organization funded by insurance companies to attend fires and other emergency incidents in order to reduce or prevent resultant damage.
Long Ladder: Sixty-foot fixed-base predecessor of the turn-table ladder.
M2FN: radio call sign for channel four on the London Fire Brigade network, used by Soho Fire Station.
Make up: an incident in which the number of appliances has been made up to a greater number than the initial attendance e.g. a four pump fire.
Make up the gear: clear up and re-stow all equipment used.
MFB: Metropolitan Fire Brigade, formed 1 January 1866, superseding the London Fire Engine Establishment.
Multiple Call: an incident about which the fire control room receives more than four or five separate calls in a short space of time, indicating beyond all doubt that the incident the appliances are proceeding to is genuine.
National Fire Service: formed in 1941 to provide the whole country with a unified fire force (disbanded in 1947).
On the bell: driving fast to a fire call ringing the bell, using two-tone or electronic sirens and blue flashing lights.
On the run: an expression dating from early Victorian fire stations when the fire engine stood ready for action on a sloping section of floor designed to assist the speedy harnessing of the horses to the appliance. Today anything in the fire brigade said to be 'on the run' is in good working order. Hence 'off the run' means out of order.
Ordering: a mobilizing message for a fire appliance or senior officer is known as an ordering (and the control officer on the radio will begin such a message with the command order, e.g. 'Alpha 242 order your pump to fire, 151 Oxford Street, W.1.').
Persons Reported: priority message to indicate that people are reported to be trapped in a fire. An ambulance is ordered to a fire as soon as this message is received in control.
Proto Breathing Apparatus: particular make of BA used prior to the introduction of compressed air cylinders. The proto sets required the user to wear distinctive nose clips and eye goggles.
Pump Escape: dual-purpose pumping appliance that carried the fifty-foot wheeled escape rescue ladder.
Pump Ladder: dual-purpose pumping appliance presently in service and carrying the 1–3–5 aluminium ladder.
Pyromaniac: person who repeatedly fails to resist the temptation to light fires.
Red Riders: the regular firemen during the Second World War who rode the red-liveried fire appliances.
RT: radio telephone, carried by all fire appliances in the London Fire Brigade fleet.
Rubbish Loonies: the firemen's name for arsonists and pyromaniacs who prowl around the streets of Soho setting fire to piles of rubbish.
Running Call: emergency call made by a member of the public, either by flagging down a fire appliance in the street or by picking up the telephone on the outside wall of the fire station.
Seat of the fire: the fire's point of origin: the effort is made to attack the seat of the fire whenever possible.
Shout: An emergency fire call.
Smoke Issuing: fire brigade description of smoke seen emanating from a building.
Special service call: any non-fire emergency call.
Spit: tea in LFB lingo.
Spreaders: hydraulic equipment for prising apart metal.
Starting up: to turn on a breathing-apparatus set.
Station Commander: Assistant Divisional Officer who commands four watches at one fire station.
Station Officer: rank of officer in charge of a watch at a station.
Stiff: a dead body in LFB lingo.
Sub-Station: satellite fire station established at the outbreak of the Second World War in order to provide extensive fire cover in central London.
Tally Board: noticeboard used by the Breathing Apparatus Control fire-fighter to monitor the number of BA wearers operating within a fire situation, the time each one 'started up', and the time they are due to be back out of the building.
TL: turn-table ladder, usually a hundred feet in length, but earlier models were eighty feet.
Trailer Pumps: pumps used during

the Second World War, usually pulled behind converted London taxis.
Turncocks: water company engineers responsible for providing water under adequate pressure to the pumps at the scene of a fire.
Turning over and cutting away: once the fire is extinguished all remaining smouldering pockets must be accessed to ensure that the fire cannot re-ignite at a later date. Much time is spent sifting through charred, and cutting away walls, floors, beams, etc.
Turn out: mobilize to an emergency fire call.
Watchroom: the nerve centre of the fire station, where the log book is kept, where emergency calls are received via teleprinter and where originally firemen would keep watch twenty-four hours a day.
Wembley Control: one of three '999' fire control centres established in the 1960s (the other two were at Stratford and Croydon), replaced by the new CMC in 1990.
Working Fire or **Working Job:** fireman's term for a serious fire.
Yaffle: old fireman's expression for taking in a lot of smoke (also 'fire eater').

NICKNAMES

Soho is famous as the birthplace of great Fire Brigade nicknames. These include:

Arnie
Baldric
Baron
Basil
Beaker
Belsen
Blister
Bloater
Bones
Botty
The Bubble
Bubbles
Bucket Mouth
Bungalow Bill
Carrots
Cavity Wall
Cheesy
Cosy
Creeper
DC
DiDi Duck Duck
Dingers
Dotty
Dracula
Drone
Fatboy
Flaky
Flipper
Gerbil
Goat
Gruesome
GW the Grey Wraith
Joe the Wop
Knob
The Lazy K
Lionel
Lofty
Mad Dog
Manuel
Moosehead
New Leaf
Noddy
Pedlar
Pencil Neck
Pervy
Plank
Poorly
Pug
Raincloud
Ramjet
Ratty
Razor
Rigsby
Ron Digsworth
Silver Fox
Skippy
Slimer
Slugger
Sperm
Spond
Spreaders
Sweatbox
Tannoy
TC
Terrible Ted
Tonka
Vinegar Joe
Wally
Whacko
Wilks
Winkle
Winnie
Wogan
Yee-Haa

APPENDIX I

Firemen who gave their lives on Soho's ground during the course of service:

Senior Fireman Benjamin Cummings
31 March 1847 Denmark Street, St Giles
(Killed by falling from the roof of a house in Denmark Street while attending a chimney fire.)
Senior Fireman John Eilbeck
27 September 1859
John Street, Tottenham Court Road
(Killed when the roof fell in on him.)
4th Class Fireman Thomas Ashford
7 December 1882
Alhambra Theatre, Leicester Square
4th Class Fireman Henry Berg
Injured 7, died 28 December 1882
Alhambra Theatre, Leicester Square
4th Class Fireman Frederick Fielder
26 November 1892
Agar Street, Strand
Citation: Whilst endeavouring to enter a third floor window, Fielder, who was exhausted by his exertions in pushing the escape from Scotland Yard, was overcome by a rush of smoke and fell to the ground from the escape. He died in a few hours from the injuries sustained.
4th Class Fireman Martin Sprague
29 October 1895
New Church Court, Strand
Citation: Whilst engaged in searching people reported to have been buried in the ruins of the burning building, Sprague was crushed by the roof which collapsed. He died on the following day.
Senior Fireman LFB Thomas Curson
16 September 1940
New Cavendish Street
Auxiliary Fireman Albert Evans
16 September 1940
Great Portland Street
District Officer Leonard Tobias
Injured 16, died 17 September 1940
Great Portland Street
Auxiliary Fireman Donald Mackenzie
Injured 17, died 18 September 1940
John Lewis, Oxford Street
Auxiliary Fireman George Abrahart
18 September 1940
Rathbone Place
Auxiliary Fireman Arthur Batchelor
18 September 1940
Rathbone Place
Leading Aux Fireman Jack Bathie
18 September 1940
Rathbone Place
Leading Aux Fireman George Bowen
Injured 18, died 19 September 1940
Rathbone Place
Auxiliary Fireman Harold Gillard
18 September 1940
John Lewis, Oxford Street
Auxiliary Fireman Robert George
18 September 1940
Rathbone Place
Auxiliary Fireman Benjamin Mansbridge
18 September 1940
Rathbone Place
Auxiliary Fireman Myer Wand
18 September 1940
Rathbone Place
Auxiliary Fireman Frederick Mitchell
7 October 1940
Soho Fire Station
Station Officer William Wilson
7 October 1940
Soho Fire Station
Auxiliary Fireman Henry Davidson
9 December 1940
Wells Street
Auxiliary Fireman Stanley Randolf
17 April 1941
Tavistock Square
Auxiliary Fireman Harry Skinner
17 April 1941
Upper Woburn Place
Station Officer Fisher
20 December 1949
Covent Garden
Station Officer Fred Hawkins
11 May 1954
Langley Street, Covent Garden
Fireman Arthur Batt Rawden
11 May 1954
Langley Street, Covent Garden
Fireman Charles Gadd
23 May 1954
Langley Street, Covent Garden
Station Officer Colin Townsley
18 November 1987
King's Cross Station

Appendix II

Significant fires, rescues and incidents on Soho's ground:

2 June 1780 Riots and firing of various Catholic chapels including the Portuguese Ambassador's Chapel in Glasshouse Street.
1785 Compton and Greek Streets burned to the ground – hundreds made homeless.
Winter 1792 The Pantheon, Oxford Street – burned to the ground.
29 March 1849 Olympic Theatre, Wych Street, Strand.
15 March 1856 Covent Garden Theatre severely damaged.
28 February 1865 Savile House destroyed by fire and explosion.
14 February 1866 Laurie and Marner Ltd, Hanover Square destroyed by fire.
6 May 1868 18 Drury Court, Drury Lane – three lives saved by Abraham King.
26 September 1868 12 Castle Street, Leicester Square – six lives saved by Richard Mockett, Thomas Hatton and Robert Offord.
27 March 1870 1 Grafton Street, Soho – six lives saved by James Seargent and George Stubbings.
19 May 1872 26 Princes Street, Drury Lane – six lives saved by John Howard and John Abraham.
10 December 1877 Villiers Street, Strand – four lives saved by Samuel Goodall et al.
1 March 1878 Buckingham Street, Strand – three lives saved by Samuel Goodall.
24 November 1878 Drury Lane – four lives saved by Samuel Goodall.
17 April 1880 High Holborn – eight lives saved by Philip Reuby.
6 September 1880 Drury Lane – four lives saved by William Emanuel.
26 October 1880 Oxford Street – four lives saved by John Scott.
11 July 1881 Goodge Street – two lives saved by Thomas Lynch (both people were carried down the escape from the third floor of the house while the whole of the rest of the building was on fire).
7 December 1882 Alhambra Theatre, Leicester Square – two firemen killed.
22 December 1891 Tottenham Court Road – twenty-five pump fire in department store.
16 February 1896 Church Street, Soho – nine residents killed in tenement fire.
26 October 1924 Wardour Street, Soho – film factory destroyed with £1 million loss.
27 December 1937 HMV, Oxford Street – building burned to the ground.
12 September 1940 Burlington Arcade – destroyed by bombing and fire.
7 October 1940 St Anne's Church, Dean Street destroyed.
14 October 1940 Carlton Club, Pall Mall – two killed, most of the cabinet and Harold Macmillan in dining room below the library where the bomb exploded escaped unhurt.
14 October 1940 St James's, Piccadilly (1676 Wren church) – the verger and his wife were trapped beneath fallen masonry and died despite a twelve-hour struggle to reach and rescue them.
26 October 1940 St James's Residences, Brewer Street – five killed, forty-five injured.
8 March 1941 Café de Paris, Coventry Street – dozens killed, hundreds injured.
8 March 1941 Madrid Restaurant, Dean Street – seventeen killed.
17 April 1941 Jermyn Street – parachute mine and high explosive bombs – twenty-three casualties including seven dead.
17 April 1941 Newport Dwellings, Newport Place WC2 – parachute mine – eighty-three casualties including forty-eight dead – police had to be called in after this raid to stop looting. (Admiralty Arch was hit in the same raid.)
10 May 1941 St Clement Dane, Strand burned to the ground.
11 May 1941 Carr's Hotel, Clarges Street received a direct hit – a perilous rescue operation to free a woman pinned to her bed by debris was effected whilst water was sprayed on a wall to keep the fire at bay. Five killed, three badly injured.
30 June 1944 V1 rocket attack on Aldwych – numerous casualties.
20 December 1949 Major fire at Covent Garden, one Station Officer killed.
18 April 1952 Second major fire at Covent Garden requiring more than fifty pumps.
11 May 1954 Large fire at Poparts Warehouse, Langley Street. Partial collapse of the building killed two firemen and a Station Officer.
1961 London Press Exchange – twenty pump fire.
1961 Grand Buildings, Trafalgar Square – twenty pump fire.
1970 Great Marlborough Street – twenty pump fire.
1971 Charing Cross Hospital – twenty pump fire.
2 January 1975 Lloyd's Bank, Aldwych – fifteen pump fire.
27 August 1975 213 Tottenham Court Road – twenty pump fire.
1977 The Plough Pub, Museum Street – three children rescued.
1977 Alfred Place – thirty pump fire.
26 January 1979 Villiers House, Strand – forty pump fire.
27 April 1980 The *Old Caledonian*, Victoria Embankment – five firemen injured in flashover.
16 August 1980 Denmark Place – thirty-seven killed in arson attack.
11 June 1981 Covent Garden underground station – eight pump fire.
21 June 1981 Goodge Street underground station – twelve pump fire.
11 January 1982 Mister Byrite, 233 Oxford Street – fifteen pump fire.
27 July 1982 Civil Service Stores, Strand – twenty-five pump fire.
18 August 1982 Gerrard Street – seven people murdered in gangland attack.
2 February 1983 Old Charing Cross Hospital – multiple rescues, two killed.
1984 6 Charlotte Street – six pump fire, sixteen multiple rescues.
2 November 1984 Oxford Circus underground station – thirty pump fire, one thousand rescued.
18 November 1987 King's Cross underground station – thirty pump fire, thirty-one killed.
15 August 1989 Lower Regent Street – twenty pump fire, four TLs.
16 February 1989 Carlton House Terrace – twenty pump fire, two TLs.
12 February 1990 Savoy Theatre, Strand – ten pump fire, theatre destroyed.
31 March 1990 Grand Buildings – eight pump fire. Riots in Trafalgar Square: dozens of cars and piles of rubbish set alight by mob.
7 February 1991 Bomb explosion and mortar attack on Downing Street.
5 August 1991 IRA firebomb attack, Cambridge Circus – five people rescued down TL.
1 December 1991 IRA firebomb attacks around Tottenham Court Road.

Appendix III

Facts and Figures

1975 Total number of calls 1,129, of which 84 were malicious false alarms.
1976 Total number of calls 1,967, of which 77 were malicious false alarms.
1983 Total number of calls 3,056.
1990 Figures Total number of calls 8,210, of which 587 were malicious false alarms.

January	621
February	569
March	612
April	591
May	709
June	680
July	704
August	735
September	714
October	577
November	749
December	749

(Including 621 Property Fires, 441 Rubbish Fires, 4 Chimney Fires, 1,491 Automatic Fire Alarms, 2,627 False Alarms – Good Intent, 587 Malicious False Alarms, 940 Special Services, including shut in lifts, lock out/in and RTAs, and 3,402 TL Fire Calls.)

Fatalities	3
Rescues/led to safety	26
Firemen injured at fires	5

Make-ups 1990:

Four Pump Fires	12
Six Pump Fires	3
Eight Pump Fires	2
Ten Pump Fires	1
Total	18

APPENDIX IV

Two twenty-four hours of calls at Soho Fire Station. These extraordinary figures speak for themselves: Soho is without doubt one of the busiest fire stations in the world. In stark contrast, sometimes the station receives only two or three calls in twenty-four hours.

19 May 1990

Time	Address	Soho Appliances Attending	Type of Call
0014	Queen's Gate	TL	AFA
0053	Charing Cross Road	PL P	Malicious False Alarm
0120	Albany Street	TL	Fire
0533	Regent Street	TL	Fire
0638	Thurloe Place	TL	AFA
0708	Oxford Street	PL P TL	AC
0721	Park Street	TL	AC
0743	Great Queen Street	TL	AFA
0934	Huntley Street	P TL	AC
0940	Piccadilly	PL	AC
0951	St Johns Wood	TL	AC
1027	Royal Free Hospital	TL	AC
1039	Grosvenor Street	TL	AC
1134	Whitfield Street	PL P TL	4 Pump Fire
1233	Old Marylebone Road	TL	AC
1322	Floral Street	P	Shut in Lift
1423	Cadogan Place	TL	AFA
1438	Jermyn Street	PL P TL	AC
1457	Neal Street	PL P TL	AC
1518	Kilburn High Road	TL	Fire
1538	Avenue Road	TL	Fire
1548	King's Cross	PL P	AFA
1602	Praed Street	TL	AC
1614	Leicester Square	PL P	AFA
1622	Stoke Newington Road	TL	4 Pump Fire
1654	St Charles Square	TL	Malicious False Alarm
1716	Chichester Road	TL	Fire
1739	Shoot Up Hill	TL	AC
1751	Regent Street	PL P	Malicious False Alarm
1806	Sloane Street	TL	Fire
1850	Du Cane Road	TL	AC
2034	Canada Way	TL	Malicious False Alarm
2117	Brondesbury Road	TL	AC
2138	Haverstock Hill	TL	AC
2259	Panton Street	P	Person Locked Out
2321	Endsleigh Place	P TL	AC
2326	Ossulton Street	TL	Malicious False Alarm
2347	Marylebone Road	TL	AC

21 February 1991

Time	Address	Soho Appliances Attending	Type of Call
0000	King's Cross	TL	AFA
0059	Cornwall Gardens	TL	AFA
0111	Covent Garden	PL P TL	AC

0136	Piccadilly Circus	PL P TL	AC
0204	Seymour Place	TL	AC
0228	St Stephen's Crescent	TL	Fire
0259	Gloucester Road	TL	Fire
0317	Tottenham Court Road	PL P TL	AC
0713	Strand	PL P TL	4 Pump Fire
0810	Hammersmith Road	TL	AC
1010	Ludgate Hill	TL	AC
1016	Waterloo Station	P	AC
1021	Lower Marsh Street	PL TL	AFA
1129	Colville Road	TL	AC
1133	Soho Street	PL P TL	AC
1207	Lodge Road	TL	AC
1211	Ingestre Place	P	Shut in Lift
1224	Finchley Road	TL	Fire
1333	Rowland Hill Street	TL	AC
1404	Maple Street	TL	Chimney Fire
1458	Sloane Street	TL	AC
1528	Hamilton Place	TL	AC
1559	King's Cross Road	TL	AC
1644	Shaftesbury Avenue	P	Rubbish Fire
1724	Bryanston Street	TL	AC
1734	High Holborn	P	Flooding
1741	Brompton Road	TL	AC
1758	Hanover Street	PL TL	AC
1823	Curzon Street	PL P TL	AC
1825	Shaftesbury Avenue	TL	Person Locked In
1843	Queen Square	P TL	AC
1912	Cornwall Gardens	TL	AFA
2019	St Anne's Road	TL	Fire
2030	Exeter Street	PL	Rubbish Fire
2033	Cartwright Gardens	TL	Person Threatening to Jump
2046	Curzon Street (Soho's machines on other calls)		AC
2109	Bloomburg Street	TL	Malicious False Alarm
2157	Guilford Street	TL	AC
2322	Acton Lane	TL	AC
2334	Edgware Road	TL	AFA
2350	Leadenhall Street	TL	AC

(AC = Alarm Caused: an automatic fire alarm has actuated because of smoke e.g. from a toaster, or with good intent someone has called the fire brigade because they thought there was a fire. AFA = emergency call made in response to an automatic fire alarm fault.)

Index